Objects and Structures

on

Numbers, Infinity, and Infinite Processes

Objects and Structures

Brian Capleton

Amarilli Books

amarillibooks.co.uk

Copyright © Brian Capleton 2022

First Edition 2022

The right of Brian Capleton to be identified as the
author of this work has been asserted in accordance with
the Copyright, Designs and Patents Act, 1988

ISBN 978-0-9935372-6-4

It is important that we understand numbers. But not just so we can come to understand the nature of our world. We might say they are given to us so that we can come to know the nature of this intelligence we are being here, and the nature of its relation to our world.

Our world has an objective part - material nature. For a very long time indeed we have made the mistake of presuming that the objectivity of the world lays in what we experience as *matter*. That experience is, however, a construct of brain function. The objectivity of the world does not lay in what we experience as *matter*, but rather, in *structures*. Structures of things. Things that are objects, in the sense that we take them to be distinct. Distinct from each other, and distinct from the intelligence we are using to contemplate them.

We can take whatever we like to be distinct objects, but when we do, we are looking at *structures*. *Structures* of *relations* between objects taken to be distinct. It is only in the structures of those objects that are genuinely distinct, that the truly objective facet of nature lies.

CONTENTS

CONTENTS

Preface

The Object Theory we refer to here is a schema of thought, rather than a thesis. It is essentially a suggestion of somewhere similar that we may need to explore as we move towards new horizons in the scientific knowledge of our own mind and the nature of our own intelligence.

Where we are now, in these times, we have already realised that mathematical "proofs" are relative things, rather than absolutes, and that some things must remain forever uncertain or undecidable. This knowledge was achieved through intellectual reasoning itself. Through the use of the human mind. But now, there is another dimension to this story. We can no longer be confident that mathematical reasoning is the study of something separate from the processes of mind that are being used to do the reasoning.

"Traditionally" mathematicians have loved to build towers of logic and reasoning, certain mathematical structures, in other words, "proofs", without ever considering the structures in the psychology and functioning of the minds doing the reasoning, independently of the act of reasoning itself. It is as if we have had a religious faith in the idea that we been discovering truths about something that is independent of us. Independent of the mind we are being, and the nature of our own intelligence. The methods used since the ancient Greeks and before, have succeeded in relating to the way things are, and behave, in the apparently timeless world of numbers and mathematical structures. But is this world of numbers and mathematical structures really fundamental to the phenomena of our world?

The position of modern neuroscience is that there is nothing we can think, comprehend, understand, perceive, or conceive, that is not a construct of brain function. In short, the implicit position is that the human mind, indeed the human intellect, and the mathematical intellect, is a construct of brain function. So brain function is a context even for our understanding of *mathematical reasoning*. It means that we no longer truly have the luxury of considering the psychology of our mental processes as a subject separate from the way we *do mathematics*.

Details about brain function are beyond our scope here, but it is well known that we do have a natural cognitive bias in the way we understand things. Psychologists speak of the *apperception mass*, and we can consider that to inevitably arise from the principle of neuroplasticity in brain function. We have to learn new things through the medium of what we have already learned.

We have seen this particularly in the difficulties we had in coming to terms with the facts of quantum mechanics. Our pre-conditioning led us to resist, for example, the idea that something can be in two places at once, or that until it is observed or measured, it doesn't exist except as a probability wave.

There are certain things that we must appreciate, once we know that what we are looking at, or contemplating, is in fact a construct of brain function. The first is that our brains and minds do not exist in isolation. On the contrary, they exist literally, in networks and overall as a network. That means that it is possible for whole *networks of networks* of minds to come to a consensus on something, that is literally a *network consensus*. In that, whatever structures and processes of mind and thought are underlying the reasoning taking place in any individual mind, it will be related in some way to what is inherent in the network as a whole. In a sense, we can even speak of "network mind".

Neuroplasticity in brain function plays a key role in learning, and in *mathematical* learning especially, *teaching* also plays a key role. It is through these mechanisms of teaching and learning, together with the principle of neuroplasticity, that "network mind" in the form of "schools" of understanding, can come into existence.

Not all mathematical theorising is empirically testable in the way that successful theories of physics have been. The history of set theory in its application to the concept of infinite sets is sufficient demonstration of that. So if empirical testing is not available, the question then arises, what exactly is it, that mathematicians are looking for, that is other than, or better than, just a matter of consensus? What is it that constitutes even just what is aspired to, or believed by some to *be* "truth" or "proof" in mathematics?

The position we have currently come to, collectively, is that we must show that there is consistency of theories and results, and structures of understanding, with the axioms upon which they are built. We now know that there are always axioms, and there are always things that must remain *undecidable*.

For our purposes here, where we are putting mathematics in the context of the modern neuroscience, we are going to be talking about something that we define as "genuinely objective". We are going to *define* this notion of *genuinely objective* as something that is not dependent on an individual's mind and brain function, but more importantly, *not dependent on a consensus* in networks of individual minds and brain functions.

This idea that there is something that is "genuinely objective" is by no means an assertion that there is anything that we can ever know about that is other than a construct of brain function. Rather, it is simply saying that what is "genuinely objective" is not dependent on the way any particular individual mind or networks of mind, or brain or networks of brains, is working or functioning.

In this way, the principle of the brain itself, the same principle that nature instantiates in all our brains, the *principle through which* we have our *experience of being*, and experience of mind, and thought, and indeed, mathematical reasoning and understanding, *that principle*, itself, we can take to be "genuinely objective".

Premises

All mathematical thought is in the broadest sense *object-oriented*. It is all about mathematical objects and their relations. And so it is that we are going to be talking about something that in this context, we are calling Object Theory.

This Object Theory is not some formal, mathematical theory built by using the mind without examining the way in which that mind is working. Object Theory is as much about the way structures of comprehension and thought processes are working, as it is about the things that are being thought about. As a method, it enables us to use our intellect to examine some fundamentals in a way that is potentially useful.

The first premise of Object Theory we have already just talked about, and this is the idea of what is "genuinely objective". It states that what is "genuinely objective" is not dependent on the functioning of any particular, individual mind, or network of individual minds. For the word "mind" in this, we could substitute "brain function". Whenever we talk about this, the *network* of minds or brains is by definition taken to be a *subnetwork* of the network of all human minds or brain functions.

The second premise is that what we are thinking about, when we are doing object oriented thinking, is *objects* that are *instantiated* in our model of understanding, in *structures*, in which they are related through *relations*. We will talk more about the nature of *structures* and *relations* between objects and structures quite soon.

The third premise is that there are always *relations* between distinct objects, even if we can never know what any particular *relation* is. Relations can be symbolised by two-ended arrows or something similar such as $\longleftrightarrow$. We can also represent structures without the use of arrows, just by putting symbols for objects, together.

The fourth premise is that a given object may or may not itself be considered as a structure, depending on how we are thinking about it.

Similarly, we can consider a structure as a single object if we wish to do so.

The fifth premise is that the reason *any* plurality of objects are *distinct*, and not one and the same object, even if they are perfectly identical, is because there is some *relation* between them. In this, an "object" in the context of Object Theory is not something that has to be tangible. It can be literally anything that we are considering in the mind, as an object of thought, in object oriented reasoning. In mathematics, the endeavour is that objects of thought should be representations or thoughts in our mind of things that are *genuinely objective*.

Thus, it is not possible to have two objects that are not implied to be distinct. The fact that there are two, means there is a *relation* between them. However, if we can show that a relation we are looking at does not properly pertain to the objects as we are considering them, then they may be two *representations* or *images* of one and the same object. If we believe the *relation* to properly pertain to the objects, but it does not, then it is a *false relation*.

So, for example, we might have a written number "3" that is identical to another written number "3". We write them both on a piece of paper, and don't presume any other context other than that we are referring to quantities. They are not one and the same object, simply because we have written two of them. The *relation* between these written marks or symbols is the object they are written on, the piece of paper, that they each have a different relation to (as a position on).

However that *relation* does not pertain to the quantity they stand for. It is not a relation of the *quantity objects* that the symbols refer to. Rather, it is just a relation of symbol positions. As far as concerns *the quantity object* that these symbols are referring to, then the symbols are distinct *representations* or *images* of one and the same object - the *quantity object* we call "3". This is because we have not yet specified any relation between the two quantities that the symbols refer to.

However, we may now write "3 = 3" intending as usual to refer to the quantities themselves. Now, even though the quantities are identical, they are *two distinct* instantiations of the same quantity object, because in our conception and method of symbolising here they are

Page 5

related through the "=", which is a *relation* pertaining to quantities. They are two distinct *instantiations* of the one quantity object represented by the written symbol object "3".

Also, if we write two "3" symbols and state they stand for quantities, and are *members of a set*, then the concept "set", or the *set object*, becomes a third object that is the *relation* between them, and so we are still referring to *two distinct* instantiations of the quantity object named "3".

We can compare this to the Leibniz *Identity of Indiscernibles* which essentially asserts that in order to be distinct two objects cannot be perfectly identical. In fact this is not so, as we see *genuinely objectively* in the case of the fundamental particles of physics.

Distinction can occur between perfectly identical objects because of the presence of a third object (such as for example a "space" in which the objects exist) to which the two identical objects can have different *relations*, and thus, this makes them distinct as objects.

We can also reverse this. From this, we can also see that two objects that appear to be distinct, but identical, are only genuinely distinct if there is a *relation* between them. This *relation*, itself, constitutes a third object.

All distinctions and relations then break down by an infinite recursive process resulting in an infinity of relations between relations. A fuller explication of this is included in *Structures and Morphisms*. Thus whenever we consider a set of objects and their relations, whether in physics or pure mathematics, we are contemplating a partial structure arising from an *infinity of relations*, in which some parts are considered as objects, and others are considered as the relations between those objects. Which is which is just a matter of the way we understand things, or our paradigm of understanding.

So distinct objects including mathematical objects, always exist in *structures*, which are essentially *networks of relations*, in which the *objects* are the nodes (vertices), and the edges (the links between the nodes) of the network are the *relations* between the objects.

A consequence of the fifth premise is that the relation between two instances of identical objects such as A $\longleftrightarrow$ A, where there is *no* object distinct from A as the relation between them, is the same as

$$A \longleftrightarrow A \longleftrightarrow A$$

which in turn is the same as

$$A \longleftrightarrow A \longleftrightarrow A \longleftrightarrow A \longleftrightarrow A$$

and so on *ad infinitum*.

This is a construct that is identical to A. For example, the proposition of infinite *otherwise unrelated* "3" number objects, is no different to the proposition of a single instantiation of the "3' number object. If we write numerous "3"s on a page, however, the page itself is not a "3", so the meaning is numerous otherwise unrelated "3"s, all related through the page surface.

If we *do* instantiate

$$A \longleftrightarrow A \longleftrightarrow A \longleftrightarrow A \longleftrightarrow A$$

and then go ahead and continue replacing the relations with further A objects, *ad infinitum*, which we are perfectly at liberty to theorise about, then we have instantiated an infinite recursion, or *infinite iteration process* or IIP. In other words, any object by itself *can be considered as as* an IIP. This turns out to be important in relation to *numbers*. It means that any number *per se* can be considered as an IIP, and in fact real and complex numbers, as we shall be seeing, *are produced* from IIPs.

The sixth premise is that a *structure*, considered as a network of objects, is *fully connected* in the graph theory sense. But it is not the case that we need to know about, or refer to, all the edges or connections or *relations* between objects, in order to be talking about something meaningful that we can understand. Generally speaking, we would be treating the *structure* as though only some edges exist as *relations* between the objects. Nonetheless, the very fact that an object exists in the structure, means that it has some relation to every other object in the structure.

<u>The seventh premise</u> is that objects and structures can undergo *morphisms* into other objects and structures.

However, here we move onto a more advanced form of speculation, which appears in the accompanying volume *Structures and Morphisms*.

Part 1
Objects and Structures

1: First of All

I n science and mathematics we are dealing with things, concept-structures, systems, phenomena, and so on, all of which are considered objectively. That means they are treated as separate from the mind, intelligence, and brain function, that is being used to comprehend and to do the study.

The position of modern neuroscience is nonetheless that everything we can know, think, understand, comprehend, perceive, and conceive, is all a construct of brain function. It is that the mind and intelligence we are being, applied to scientific and mathematical study, is a construct of brain function.

No one can deny that objective (object-oriented) study as it is carried out in science and mathematics, has been and is successful. It has led to the technologies through which we have changed our world. So even if we want to say that everything that we understand, perceive, and conceive, about our world, and indeed, about mathematical structures, comes down to structures in brain function, we have no need to think of questioning the principle of objectivity. Objectivity works.

In physics there are concepts such as "energy", for example, that cannot be understood or explained in isolation from the networks of concepts in which they appear. In mathematics we find a similar thing, in terms of networks of axioms. The axioms of mathematics, and the fundamental concepts of physics, are taken to be objective, and treated as objective facts. That means, they are taken to be, and treated as, separate from, and independent of, the mind and intelligence that is being used to comprehend them.

Historically, this has appeared as what has been called *Cartesian duality*. That is, the distinction, as asserted by Descartes, between mind and matter. Or in the case of pure mathematics, between mind, and the mathematical structures that the mind is contemplating.

Today, however, the assertion of this duality is not consistent with the fundamental premise of modern neuroscience. That premise is that there is no distinction between mind and matter, as far as mind appearing in human beings, through brain function, is concerned. Any other kind of mind that we might want to propose, is not be something that mainstream neuroscience comments on.

Given this situation, we can reasonably conjecture that the structures of understanding that we work with in mathematics and science, have some kind of correlation to the complex structures of functioning that we see in the brain.

What we refer to as "Object Theory" in this book is not in itself, in its current "incarnation", a constructive theory for making predictions about the behaviour of natural phenomena. Rather, it is a tool of deconstruction, not for the sake of deconstruction itself, as in "deconstructionism", but with the positive purpose of a deeper understanding.

Object Theory is about what it means to be "objective" in our thinking, in the context of neuroscience, and brain function. It is not explicitly about brain function, but about the relation between objective thinking and natural phenomena. Natural phenomena includes the phenomenon of the brain itself, through which arises our intelligence and mind.

The matter of what is "objective" in this context, is a *defined part* of Object Theory, that we will come to in due course. As far as Object Theory is concerned, we are only concerned with what is being treated as, and taken to be, objective, in our comprehension, theorising, and reasoning. The simple *presumption* that something is objective, in mathematics and science, is insufficient for it to qualify as an *object*, as far as Object Theory is concerned.

If we allow that what we experience as mind and understanding, and conceptualising and reasoning, arises through the unimaginable complexity of brain function, then what is going on in Object Theory, should be more clearly grasped.

We are not going back to some supposed "first principles" as if hypercomplex structures are already fully understood. Collectively, we are nowhere near ready to embark on this understanding. It is unlikely that we can begin on that path, until powerful quantum computing is widely available.

However, this does not matter, because we are already where we are, in mathematical and scientific understanding, and in using the principle of *studying things objectively*. It is true that we are using our mind to do this understanding, without understanding this mind that is doing the understanding. But nonetheless, we are sufficiently far down the path of knowledge, using the principle of objectivity, to take a high level, generalised view of what it is that we are doing.

This is what Object Theory does. It is not a theory about things that *the theory itself* takes to be other than a construct of brain function. The theory is not positioned on the assumption that anything we can know or experience is other than a construct of brain function

It is not a theory about things that *the theory itself* takes to be other than the processes of mind we are employing in understanding natural or mathematical phenomena. So in a sense it is about the psychology of intellectual reasoning. It is about the underlying processes of reasoning employed when we understand things in science and mathematics, objectively.

Object Theory is a method for representing the concept structures in what we are doing when we are reasoning objectively, and what we are looking at, in the things that we contemplate in science and mathematics. Object Theory is one way of approaching thought processes in object oriented thinking, that does not presume we are able to distinguish between the mind, what the mind is thinking about, and tacit processes that lay behind the way we reason.

Here, we are focusing mainly on mathematics. And specifically, we are focusing on some very basic, fundamental principles and objects in mathematics, such as numbers, iteration processes, and the concept of infinity.

2: Objects and the Objective

The first thing is the idea or concept of an "object". We are now about to consider this in a way that is not lacking in parallels to the concept of an "object" as it arises in object oriented computer programming.

In the context of Object Theory an "object" is an object in the "field of thought" that we are using when we are applying our intelligence to something, "objectively". So this can be both some material phenomenon, or material thing, and nonmaterial things, such numbers and symbols and mathematical structures. Sometimes one hears talk of "numbers" versus "concrete numbers". Object Theory gets underneath this, as it were, into the psychology behind it.

Something doesn't have to be "fixed" and unchanging, it doesn't have to have any further properties or qualities, other than that it is *considered as* an "object", in order to qualify as an "object" in the context of Object Theory. So what we are saying is that the very idea or concept of an "object" in our "field of thought" is a primary position that already exists in our *modus operandi* of thinking. It is a primary position that our *modus operandi* of thinking is *already taking*, even before we refer to any particular or specific object, that has properties or qualities, or identifying features, that we can talk about. This fundamental, underlying idea or concept of "object" is what exist in our thinking psychology before we even begin thinking "objectively".

This is the "top level" *object* in Object Theory, that can be *instantiated* as many times as we like, in as many diverse "incarnations", with various properties or qualities, or identifying features, as we like.

So let's look at an example. Let us take the number "3". If we are being "objective", then the first thing that happens in our mind when we think of the number "3", is represented in Object Theory, as the *instantiation* of an *object*.

But when we think of "3", what we instantiate in our mind is an *object* that comes ready-furnished with the property "3", and our whole *apperception mass* associated with that, which also acts as an identifier of the object. The Object Theory analysis of this is that this object "3" actually consists of the primary, abstract object: *object*, that has been then furnished with the property "3".

Now this property "3" is itself an object that already exists elsewhere in our mind, comprehension, intelligence, or correlate in brain function. It is something we have previously learned about. However, as an individual, when we think of it, the thought of it only happens "in our own brain", and so it is subjective. It becomes part of objective thought, by our first (unconsciously) instantiating the idea of "object" - which is the fundamental and primary abstract object: *object* - and then applying the subjectively reconstructed thought-object "3" to the primary *object*.

This is how Object Theory treats all objects of thought, in "objective" thinking. All our objective thinking relies on what we have learned, and on psychological *apperception masses*. Everything we think or imagine, in objective thinking processes, and the process of objective reasoning itself, is a *reconstruction* in brain function. It happens only in our own, individual brain. However, what we discover through the objective pursuit then becomes understanding some aspect of our world that is not dependent on our individual brain function - the objective part of the world's nature that science studies. In this way, we make progress in science.

This touches upon a question in neuroscience that in the mainstream is currently regarded as undecidable. That question is about the nature of the world "external" to brain function. It is sometimes expressed as the question "*Is there* a world external to brain function"?

Given that there is nothing we experience, know, or understand, nothing we can conceive or think of, or perceive, that is not a construct of brain function, and in some way arising through brain function, then what is the true nature of the world that we *conceive* as being "external" to our experience and perception of it? When all we actually experience, think, and know, is just a construct of brain function?

In other books I address this, but it is beyond our remit here. What we undeniably know, is that the material world we know and experience - which in our experience and knowing of it is a construct of brain function - *does have an objective aspect to its nature*. That is the aspect of the world that science studies, and the part that mathematical structures, as we have discovered and understand them, are successful in "modelling". What this means, we will come to in just a moment, in the Object Theory definition of "genuinely objective".

What happens in the case of pure mathematics? There is a part of pure mathematics that mathematicians in general regard as being "objectively true". But we have now reached the point in the evolution of a mathematical understanding where we know that by "objectively true", what we really mean is that whatever it is we are looking at, is consistent with the axioms of that particular mathematical model. We know that we cannot go further than that, into some "absolute truth". Model Theory is itself something that deals with this.

As far as Object Theory is concerned, what is an *object*, and therefore "objective", is whatever we are regarding as such. This might superficially look like a weakness in the theory, but actually, it is its strength. This comes down to what we were just talking about in relation to the construct of brain function that constitutes our mind and experience of our world.

Object Theory itself makes no presumptions about the answer to the question we posed, about the relation of the objective world in which we live, to brain function. Rather, Object Theory simply identifies a distinction between what is considered to be "objective", just because that is the way we are thinking about it, and what it *defines* as "genuinely objective". What it defines as "genuinely objective" is that which is *not dependent on any individual brain function, or, indeed, any subnetworks of individual brain function*. (It can still be dependent on the whole network of all human brain function).

This definition takes what we have already succeeded in establishing in science and mathematics as the *objective aspect* of the nature of our world, which is clearly not dependent on individual brain function or networks of individual brain functions, and relates that to where we are at, in our current knowledge of the human brain.

Page 16

It recognises that there is an aspect of human brain function through which we have collectively come to understand what is genuinely objective about our world. But it also recognises that what we consider to be "objective" or "objectively true", *may not necessarily be* "genuinely objective" in the sense that it has no dependence on anyone's particular brain, or on any subnetwork (of the entire human brain network) of brain function.

This is an important point because it does not follow that because a statement of mathematics is logically consistent with its axioms, that what it says is *genuinely objective*, according to the definition of *genuinely objective* set out in Object Theory.

The prime example of this is theories about infinity. The Cantorian idea about the infinities of the continuum and the reals, which was initially rejected, and later enjoyed acceptance, has more recently been shown through Model Theory to be false. The longevity of its acceptance was clearly based on network consensus. The network consensus considered its position to be an understanding of something objectively true. But what is objectively true, isn't necessarily *genuinely objective*.

3: Instances, Distinctions, and Structures

A given version of an app (software) is an object, and we can think about it in objective terms. It is a good example of an object, in the sense of Object Theory. It is also *genuinely objective* by the criterion of Object Theory because it doesn't depend for its existence and nature on anyone's *individual* mind or brain function, or even on a *subnetwork* of individual minds or brains.

Each installation of the app on a computer or mobile device is an *instantiation* of that object. Each installation or instantiation can be a little different, depending on the hardware, but the object is the same. Alternatively, the instantiations can be identical. They are nonetheless distinct. An object can be instantiated multiple times, and the instantiations are distinct, even if they are identical.

So objects can be identical *and* distinct. An example from material phenomena is the electron. If we are considering two objects that are identical, but are not distinct, then the objects we are considering are not *instantiations*, but two representations, or images, of the same one object.

New York, for example, is an object that also happens to be genuinely objective. A photograph of New York is not another *instantiation* of New York. It is just a representation, or image of New York. Similarly, when we write, for example, "3", the symbol that we write is a representation or image of an object that it stands for - a *quantity* object. So we can write "3, 3, 3," with no other context, and all we have actually instantiated is three identical instances of the symbol "3", which as instances of the symbol "3", are distinct. They are only distinct, however, because they occupy different positions on the screen or paper. In other words, there is the presence of another object in order for them to be distinct.

What happens if we now want to be talking about the *quantities* that the symbols stand for, rather than the symbols themselves? One way we can do this is to give them a context. Which, whatever it is, is the presence of another object.

For example, we can write "3 = 3". In mathematics we may call this an identity. Some people might call it an equation. There is a context of agreement here, when we see this in mathematics, that what we are talking about is the quantities that the symbols stand for. But many people *think* in terms of the symbols. They build mental structures of understanding based on the symbols themselves. And then there is the tendency even to think that written symbols that we call "numbers" are special, or have metaphysical significance.

In fact, we use the word "numbers" to stand for both symbols that somewhere along the line we have invented, and the quantities that they stand for. It is *quantities* that are primary, and genuinely objective. If this is what we mean by "numbers" then we are indeed talking about something that is genuinely objective.

However, if we are talking about the symbols, then they too, are indeed genuinely objective, as ink on paper or pixels on a screen, but the actual meaning in the form and nature of the symbol, is not genuinely objective. It doesn't exist independently of the network of brains that uphold its existence through the invention of the symbols, and through subsequent teaching and learning.

So let's now assume that what we are talking about when we write "3 = 3" is the genuinely objective *quantities* that the symbols refer to. Through the use of the symbols that we call "numbers", we have instantiated two identical instances of the quantity object "3".

Notice that once again, now that we are talking about not merely the symbols on the paper but what they stand for, that the distinction between the two identical objects 3 and 3 that we have instantiated, is accompanied by the presence of a third object: the equals sign. The whole object "3 = 3" (not the symbols but what they stand for) is a *structure*. It is an example of what Object Theory calls a *structure*. It is a structure of objects, specifically, two distinct "3" objects, that are *related* by the "=" object.

The high-level generalisation of a *structure*, in Object Theory, is essentially a network of objects, in which the edges of the network (the links between the objects) are *relations* between the objects.

If we just write identical symbols standing for identical objects, but create no context by relating them to some third object (or more), such as an "=", or declaring that they are "members of a set", or "members of different sets", and so forth, then we have not actually instantiated distinct objects, other than the symbols themselves. And *they* are made distinct by a third object which is for example the paper or screen that they appear on, in different positions.

However as soon as we say these objects are *related* in some way, perhaps for example through the relation that one "equals" the other, or by saying that they are both members of a set, or indeed members of different sets, and so on, then we have created a structure, that creates a *relation* between the two *instantiations*. The *relation* is some other object or structure of objects.

In this principle all instantiated objects are already related, if they are distinct, and it is just a matter of whether or not we can find what those relations are. This is essentially what we are doing in science and mathematics.

Any object that we want to contemplate, is already a structure. It is a structure because it exists as a structure of brain function. We don't know what that structure is, but in Object Theory we don't need to. We just need to know that everything we contemplate objectively, is a structure of objects.

What we are saying here may not be immediately obvious. So a little more explication is in order. Our previous *modus operandi* of mathematical thought, has consisted of a presumed distinction between the mind doing the contemplation, or thoughts, and the mathematical objects and their relations, that we have been thinking about. This is inconsistent with the fundamental position of modern neuroscience. Which is that there is nothing that we can know or think, or contemplate, or comprehend, that is not in some way a construct, arising as a structure (in the Object Theory sense) of brain function.

In this way, all objects that we can think about, are already structures. Now clearly some people will want to say that the nature of the object we are thinking about, is not the same thing as, or is different from, the nature of the structure of brain function, through which we encounter it and think about it. But actually, this is precisely the argument we mentioned above, that there is a world that is other than a construct a brain function, that we are indirectly coming to know, through brain function, that is "modelling" it. A world that we can never know directly.

However, this is not a scientific fact. It is a proposition. As we already said above, the question of whether or not this is true, is objectively undecidable. It is a question about the true nature of a world presumed to be separate from brain function, that we are only coming to know, through brain function, as a construct of brain function. It is a well recognised question. It is not one that I am mentioning here for the first time.

And there is an alternative view, also now well recognised, that it is a redundant proposition. In other words, there is an alternative view that *all there is*, is a construct of brain function. It is beyond our scope here, to get into this any further, except to repeat that where we are currently at, in neuroscience, the question of the true nature of the "objective world", and whether or not it is other than a construct of brain function, is objectively *undecidable*.

So Object Theory is not directly taking a position on that. It is not addressing that question. Rather, it is simply re-understanding what we already know, in a new way that acknowledges the fundamental fact of neuroscience, and recognises that everything we are intellectually looking at and understanding, when we are being objective, consists of structures of objects (In the Object Theory sense). There are structures of objects that are the structures of understanding in our minds, through which we understand in science. What we are understanding through these structures, are structures in natural phenomena. And "tying these together", as it were, in some way, are structures in brain function.

Object Theory addresses all this in a straightforward way. It simply says that anything we are objectively considering as an object, can be accepted as an object in the context of Object Theory. It then says that two such objects are only distinct, where there is a third object that is the *relation* between them. Objects with relations between them are then *structures*, any of these objects are also structures, and we can consider any structure as a single object.

What we are doing, is endeavouring to find the relations between objects, that appear in structures. This is what we are doing in mathematics and science. And as long as we are being objective, we do succeed in doing that. But now, we are entering a new era. The era of brain science. An era in which we need to come to know what being objective really means.

In physical hard science we do it through measurement. But measurements, and data, are not enough. We could not have come to where we are now, in science, without employing the human intellect. It is there, that we need to know what being objective really means.

It is there, that we need to come to realise what genuine objectivity really is. And not to confuse it with the consideration of objects and their structures, that appear in our mind, when we are thinking objectively.

The important thing to remember is that when we are thinking objectively, what we are thinking about his objects. They are taken to be separate from the mind and brain function through which this thinking is taking place. Through communication, we can set up whole networks of minds and brains, that think about the same objects, and share structures of understanding. However, the only thing that is "genuinely objective", as far as Object Theory is concerned, or structures of objects that are not dependent on any individual mind or brain function, or subnetwork of individual minds and brain functions.

4: Mathematical Structures - A Short Discussion

Often in mathematics and physics two things can be shown to be equal through two different paths of reasoning arriving at the same conclusion. For example, we may have two distinct expressions A and B, each of which is a structure of terms, operators, and variables. When all variables are replaced with actual numbers, both A and B can be evaluated to a number.

Sometimes we calculate A through one means, and B through another means. Then if we can show that each is equal to a third object C, we have proven that A = B. This might happen because each means that we use, contains the common object C.

Algebraic calculation is full of such structures and manipulations of structures. In this example when we arrived at the A = B conclusion, A and B are two distinct objects, because they are *related* by the "=" object (not the symbol itself, but what it means), whilst their evaluations are also identical objects because they evaluate to "the same number". In Object Theory terms they are two distinct instances of the same number object (which is a quantity).

In physics and applied mathematics we use mathematical structures such as algebraic structures, as a way of understanding and building theories. In hard science, we can then, as it were, in the end, "cash in" these structures in the process of empirical testing, involving actual measurements (of quantities), which is genuinely objective, thus converting our mathematical structures into numbers. And so it is that empirical, quantitative verification, confirms a theory.

In pure mathematics, the test of genuine objectivity comes down to numbers. Even to begin with, valid algebraic structures that are expressed in *equations* are only verifiably valid if they "stand for" structures of *numbers* that are valid. We cannot, for example, validly write A = A + 1, because there is no number that A can stand for, that would make the structure "A = A + 1" consistent with what we understand such a structure to mean. An equation is a structure that is saying that two objects, one on each side of the "=" object, are instantiations of number objects that are distinct, but identical. Mathematics and physics are dominated by this type of structure, called an "equation", as their main tool.

Scientific understanding rests on showing how number objects that express quantities in natural phenomena, relate to each other through *structures*. The mathematical structures in general, that we use, typically equations, consist of objects that are "placeholders" for numbers, *related* through things that are distinct from such "placeholders", such as functions and operators. However, these other objects, such as functions and operators, in the context of equations, don't have any useful meaning to us *except* in relation to numbers, or other objects (symbols and letters) that are placeholders for numbers. Numbers that finally represent quantities in natural phenomena.

The seven base quantities defined by the ISQ are distinct objects. We understand classical material phenomena using structures of these objects. In quantum mechanics there is a different set of distinct objects that form the structures through which we understand the nature of the mechanics. There is also currently no overriding theory that can relate quantum mechanics to classical scale mechanics in a satisfactory way.

Whilst the "changeover" from quantum mechanics to classical mechanics is sometimes described within the domain of quantum theory as the "collapse of the wave function" or "state vector reduction", and so forth, what it comes down to is that quantum phenomena can exist in a superposition of states, whilst classical scale phenomena do not.

What has actually happened is that we have found that the mathematical structures that describe quantum phenomena, include these superpositions. The principle of superposition certainly occurs in phenomena at the classical scale, but an object at the classical scale cannot be in a superposition of states in the same way that an object at the quantum scale can. For example, an electron (an object at the quantum scale) can be in a superposition of positions in space (more than one place at a time), whilst classical scale objects cannot.

We were surprised to find that this is true of quantum objects, without really understanding *why* it is that it is not true of classical scale objects.

At the quantum scale, the location of the electron can be a superposition - two places at one point in time. At the classical scale a billiard ball can only be in two different places at two different points in time. So the principle of superposition is an important clue. Not, specifically, just this superposition. But superposition in general, as a structure. In superposition objects are combined to make one structure, one new object, but in such a way that the structure can still be considered as consisting of distinct objects, the objects that have been superposed to make the structure.

Part 2
The Brian, Mathematics and Infinity

5: Basics

What is called the "set" of natural numbers consists of distinct objects. So natural numbers are a *structure* in the context of Object Theory. That means, that they are not merely a set, but a *network* or *graph*.

They are a fully connected graph, in the context of graph theory, because every natural number has a relation (an edge in the network) to every other natural number.

When we are talking about "numbers", we have to be careful about the psycholinguistics. That is, the words "number" or "numbers" have more than one meaning. "Number" can refer to the symbol or string of symbols that we call a "number". But it may also refer to what that "number" symbol stands for. It is, after all, only a symbol. What it stands for, is a quantity.

When we make a statement such as "the natural numbers are countable", in what way is the word "numbers" being used? We all know that "natural numbers" refers to the numbers 0, 1, 2, 3,... And so on. But are we talking about the symbols, or the quantities that they stand for? You might say "both". Informally, they are often referred to as "the numbers we count with".

If we talk about *countable* and *uncountable*, in what way are we using the word "countable" here? If we are talking about the symbols, then they are "countable" *by definition*. Because they are in fact, themselves, part of the process of *counting*. They are the resultant of the process of counting. The process of counting is always an iterative process that outputs a resultant element on each iteration of the process. Each resultant element is a number symbol, or string of symbols that in itself constitutes a number symbol. It is also the case that each "number" symbol stands for a specific quantity, *by definition*.

What we are speaking of, here, as "a number", is either actually a symbol, or a quantity, the quantity that the symbol stands for. If we are also thinking, either consciously or subconsciously, of a third meaning, namely, a "number" as something transcendental to this, or even perhaps metaphysical, and so on, then we are introducing a third thing into our thought structure, that we are not actually explicitly dealing with. If we continue to think with this third object in our thought structures, this idea of a "number" as something other than either the symbol or the quantity it stands for, then we are entering the realm of psycholinguistics.

If we steer clear of this, then what we are dealing with, is symbols, that are the "output" or resultant elements of an iterative process called "counting", and specific quantities, to which those elements are applied as "labels".

When we call specific quantities "numbers", that is a result of our intelligence and its capacity to count specific quantities, and to understand counting. We invent the idea of "numbers" in the intelligence that we are being. And then, interpreting our world through our intelligence, we say that these "numbers" exist, as part of the way we understand quantities and their relations.

Now let's consider what we call the "real numbers", rather than the "natural numbers". The "real numbers" are all numbers that don't involve the imaginary square root of minus one. They can be any such number, whether a natural or "whole number", or a number that has a decimal point and numbers after the decimal point. Either way, as a symbol, they are symbols that stand for quantities. Quantities themselves don't have to be "whole number" quantities or rational. The ratio of a circle circumference to it diameter is still a quantity.

But what are the quantities that the real number symbols stand for? Are they "real numbers" too? Just as is the case with the "natural numbers", the quantities we are talking about now, are "real numbers" if we choose to call them so, thus using the term "real number" to refer to both the symbol, and the quantity that the symbol stands for. These are not matters of something separate from our intelligence and thought processes, they are matter *of* what is going on in our intelligence and thought processes. They are a matter of objects and their relations - *structures*. We may think of a number as an object precisely because it is objective. We may think of a chemical structure as an object precisely because it is objective. These things are objective, but they are also objects and structures in thought processes. They are objects and structures in the stuff of the mind and intelligence that we are using and being. A little later, we will talk about what "genuinely objective" means.

Again, here in the "real numbers" situation we have symbols standing for quantities, symbols that are also the resultant of an iterative counting process. They are that, even when we are applying them to a continuum. However, in the case of the "real numbers", the iterative counting process is a little different to the iterative counting process used for "natural numbers".

Number Generating Processes

It is helpful if we simply refer to the iteration process as a "machine". Consider the "natural number" generating machine. Here, we are using the word "number" to refer to a symbol. In order for the "machine" to produce its "output" or *resultant* symbols correctly, it has to produce resultants that are unique, that is, they are all distinct, and refer to specific quantities.

Now as we all know there is a question of *number bases*. The smallest number base is the unary number base, in which the machine's output utilises only one symbol in generating its output symbols. The output symbol in unary is either this one symbol, or the repeat of the symbol as many times as necessary. What we call "zero" is represented in unary, as the simple absence of any symbol. And then the "numbers" proceed as, for example, 1, 11, 111, 1111, and so on. The same way that the proverbial person locked up in a prison cell, marks off the days on the wall. Strictly speaking, any symbol will do, and the symbols can even be different. They don't have to be the symbol "1". There is nonetheless an *iteration* going on.

In the case of the incarcerated person, the output is the marks on the wall. If the person makes one mark per day, then the iteration process also involves the rotation of the Earth around its access. So because the isolation process is itself something that takes place in time, it is a measurement of a quantity of time.

However, although we tend to think of *processes* in this way, as "happening in time", mathematically, a process doesn't have to be a process *in time*. Any iteration function is a process, and one that doesn't necessarily imply that it must happen in time. The famous Mandelbrot Set fractal is the result of a process - an iteration function - and it takes time draw in practice, but as a mathematical structure it exists complete. It its not a function of time, and the time parameter is not part of the function. Nevertheless, the iteration function is a process.

We are all probably especially familiar with the binary number base, and the number base 10, and hexadecimal. Essentially, what happens in any number base, is the same thing - there is an iterative process. And it's not one that happens in time, unless you use it to measure time.

You could think of the process literally like a "traditional" clockwork counter with wheels, and numbers (number symbols) on the wheels. The unary number base has infinitely many wheels available, side-by-side. They are not, however, wheels that actually turn. The available wheels with the single symbol showing, are simply introduced to the right of the existing ones, on each new iteration of the machine.

In the other number bases, there are infinitely many wheels already lined up in the machine, and as any wheel *completes* a rotation, it also turns the wheel to its left by one symbol. The iteration cycle of the machine consists of a cycle of rotation of the wheel to the far right. The string of symbols across all the showing wheels, is then the output "number".

There is also an infinite number base, with one theoretical wheel, and with infinitely many number symbols on it. An iteration of the machine then consists only of rotating the wheel by one number symbol. Essentially each iteration is a cycle of a process that keeps producing unique symbols *ad infinitum*.

Bijection

It is of course not the case that counting to number base 10 is the only valid way to count. We could in principle also count using the infinite number base. When standard Cantorian theory and its derivatives talk about "bijection" or "one-to-one correspondence" between numbers, what is actually being talked about as a single "correspondence" or "bijection" is actually a *relation* between the two numbers in question. As long as this relation is consistent throughout a "set" then there is said to be one-to-one correspondence. Usually, it turns out that this is just a technical euphemism for a ratio or equality.

We don't actually need bijection (one-to-one correspondence) in order to *count* quantities or number symbols. The suggestion that we do, is already loaded with cognitive bias from the idea of *sets*. In fact, it is not mathematically true, but rather, a misdirection. The presentation of the theory of bijection as a way to "prove" that the infinity of the reals is larger than the infinity of the naturals, is a form of mentalism, or psychological illusion.

It seems to stand up, and its consequential conclusion that there are different "sizes" of infinity, seems reasonable, if we are both not very careful about the language we are using, and we simply accept the initial proposition that bijection (or one-to-one correspondence) is necessary in order to count.

Many intelligent mathematicians have in the past been convinced by this, for some considerable time, despite the open question about it known as *the continuum hypothesis*. However, recently (2017) the Hausdorff Medal was awarded to mathematicians Malliaris and Shelah, essentially for proving that the infinities of the natural numbers and the real numbers, are in fact the same "size". The work came from the study of the "problem" of the "sizes" of these infinities, in the context of the Model Theory view of theories and their complexities.

Where does that leave the Cantorian theory argument that the infinity of reals is larger than the infinity of naturals? What can we learn from it? Actually, quite a lot. Rather than just abandoning it, what we need to learn, much more importantly than that, is why anyone was (or perhaps still is) taken in by it. Because that is not just a question of mathematics and mathematical theory. It's not even a question of metamathematics. It is a matter of psychology.

6: On the Relation Between the Mind, the Brain, and Mathematics

Sometimes what becomes "established" in mathematics is just a matter of consensus. We can see this very clearly in the story of the Cantorian argument. Cantorian theory was actually rejected by the network of mathematicians who were alive at the time when Cantor proposed it. It was only much later, that it became "received wisdom" in the field of mathematics, for many mathematicians. It's not that the process consensus doesn't have a valid function, even when it is mistaken. But we need to understand what is really going on, in the progress of mathematical knowledge.

The most important point of the whole story here, does indeed concern *networks*, and the human mind. Firstly, consensus involves the network of mathematicians, or the network of the "mathematics community", if you like. Secondly, the work of Malliaris and Shelah involves model theory, and that, essentially, can be viewed as being about the network of the axioms of mathematical theories. Thirdly, all of this only arises in the first place, in the context of another network - the neural network of the human brain.

Mathematical theories and their axioms, as studied by model theory, are studied as objects, but also, in our comprehension of them they are literally organised thought structures or mind structures. However, at the current time, mathematicians like to think about mathematical problems, using the mind, without otherwise considering the nature of the mind itself. When we do this, we might tend to also consider that the objects we are thinking about, if they are not about material phenomena, are transcendental in some way, precisely because we are already considering them as *separate from the mind we are using* in order to do the study.

Neuromathematics does the opposite. It says that the thought structures in use are inseparable from the object structures that are being thought about and studied. It recognises that the whole affair is taking place as a construct of brain function. And so when model theory is looking to understand things from the point of view of mathematical "models", the "models" it is looking at are actually also manifestations of "patterns" of brain function. We have no idea how one maps to the other, or even if the concept of "mapping" is the right concept for the correspondence. But there most certainly is a correspondence between brain function, and the models that we are comprehending, in the mind.

Neuromathematics recognises that it is all a construct a brain function, but this in no way implies that valid mathematics and mathematical structures, are not genuinely objective. They are not at all dependent on anyone's individual, personal mind. They are not dependent on the functioning of any individual brain. But they are, nonetheless, in our comprehending of them, a construct of brain function.

Object Theory

Object Theory addresses this situation at the highest possible level of abstraction, recognising that valid mathematical structures are still objective, even though they correspond to constructs of brain function, and hence, are perceived and conceived and apprehended, as objective thought structures.

In order for something to be objective, in order for us to be considering it in that way, we must consider it in terms of *objects* whose workings and nature (structures) are not *dependent* on the way any individual mind (or subnetwork of minds) is comprehending these objects.

Object Theory deals with this concept of an object, at the highest, most abstract level. This is very similar to the way objects occur in object-oriented computer programming. Obviously in the context of computer programming the object appears as what could be thought of as a "code module". So there, it is not truly abstract, even though in object-oriented programming you can instantiate an object that has no properties or methods, which in a sense, is a completely abstract object.

In Object Theory, the only "property" that an object has, in its purely abstract form, is that it is an object. And that, of course, is self-referencing. So what does that actually mean? It means that as we said, it is regarded as separate from, and independent of, the mind that is conceiving and thinking about it. That's what makes it an object, rather than the subject or the mind, that or who is thinking about it.

Note that I said "regarded as". In actuality, whatever we conceive or think about, even as an object, is of course part of the mind, and part of the construct of brain function, that constitutes the intelligence and thought that is dealing with the object.

There is an underlying thesis or conjecture in Object Theory, that if it were to be sufficiently explored, there will be a point in its exploration where it demonstrates that what we are exploring in mathematics and science, is the structure of the mind that is doing the exploring.

So at the highest level of abstraction is the concept of the object. In science and mathematics, when we conceive something, we are first of all invariably conceiving an object, in the highest abstract sense, even if we are not doing so consciously. And then, that concept "object" can have associated with it, other concepts, as properties or processes. And this then becomes the object that we recognise and name as the thing we are thinking about or considering. Whether that may be a quantum particle, an atom, a chemical reaction, or a brain, and so on. Even something like an idea can be an object, if we conceive it as such, and think about it in that way.

Later, we will be talking about the concept of the "continuum" as an object, and the concept of "the real numbers" as a concept, and the relation between these two objects.

It doesn't necessarily follow that an object in Object Theory will successfully represent any objective phenomena in the material world. After all, a "bird-pig with flight" is an object, if you want it to be, in the sense of Object Theory. Object Theory does not in itself make any stipulation about what can, or cannot be, an object. Essentially, anything at all, can be an object. What about the "bird-pig with flight"? It's also a structure consisting of the object bird, the object pig, and the object flight. One way in which all those three objects are *related*, as a *structure*, is that they are all components of the object

Page 37

"bird-pig". Here we can see the three essential elements of Object Theory, which are objects, relations, and structures, in action.

Of course, this particular object is a *chimera*, a thing that we can conceive of, but that doesn't have a counterpart in the objective, material world. Nevertheless, even with a chimera, we can develop a fully fledged theory about it as a structure, and depending on what it is, we may be able to use some kind of engineering to possibly even bring it into material existence, or at the very least bring some kind of representation of it into material existence. Complex chimeras of this kind exist in the form of characters and places and situations, in computer games, for example.

So an object is anything we are thinking about, in the mind, when we are thinking about it as if it is separate from the thinking. In actuality, this separation is not true, because it is still part of the *process* of mind. Later, we will be looking at the object "process" in its own right.

The Genuinely Objective

In Object Theory the fact that something is an object doesn't mean that it is *genuinely objective*. Here, just to reiterate again, to say something is *genuinely objective*, as we defined it at the beginning, means that it is not something that is dependent on anyone's individual mind, or even on networks of individual minds. However, we exclude from this the entire network of all human minds.

So, valid mathematical structures are genuinely objective in that they are not dependent on anyone's individual mind, or on the working of some network of minds, but this does not mean that our comprehension of mathematical structures is separate from the *process* of mind.

There is nothing in this that asserts that something that is genuinely objective, either is, or is not, a construct a brain function. However, inasmuch that we comprehend it, it is part of the process of mind, and neuroscience certainly considers that to be a construct of brain function.

The premises of Object Theory are completely consistent with the neuroscientific fact that there is nothing that we can perceive or understand, that is not in some way a construct of brain function.

To make the approach we are making, in this way, using Object Theory, we don't need to first understand "what mind it is", or how biological brain function translates to it. All we need to do, is have a way of showing the *processes* of thought that we are engaged in, in the way we understand mathematical structures and scientific theories. Whilst that may seem a formidable or impossible task, it can potentially be characterised through Object Theory.

What is genuinely objective according to Object Theory, about mathematics, isn't necessarily intuitively obvious. Mathematicians have always gone to great lengths to "prove" mathematical things. Some of these things turn out to be genuinely objective, whilst others do not. Rather, it turns out that a "proof" is actually just the demonstration that the thesis is consistent with the axioms of the mathematical model it is a part of. Some things, as we now know, must remain formally undecidable.

Also, some things are not what they appear to be, from the names that they are given. For example, imaginary numbers. They were named "imaginary" for a reason. They are clearly something that arises in the intellect and mind that we are using in order to contemplate mathematical structures.

However, as structures operating in the way natural phenomena works, natural phenomena that is *genuinely objective*, "imaginary numbers" *are genuinely objective*. Whatever they are, in the way human brain function works, they are an aspect of that function that is not dependent on any individual brain function, or network of brain functions. However, they nonetheless exist in the first place as a construct of human brain function.

7: Structures and Infinity

I t must be remembered that in Object Theory, anything we want to designate as an "object" we can designate as such. Any "thing" that we want to talk about, we can, if we want to, treat as an object in the sense that it is *taken to be* separate from the mind that is thinking about it. It can itself, and most likely is, a structure of further objects. Or, alternatively, we can consider any structure as a single object.

In mathematics, infinity is an object. But we can also regard it as a structure. Of course, if we just think "infinity" we don't necessarily see any structure in that. It appears to be an object in the mind, without structure. But whenever we use the concept of infinity in mathematics, it will be associated and related to other concepts, and will be part of a structure that includes other objects.

We are going to get away from this idea of a "set", which is an object, and from all its associated structures that we will tend to fall into being preoccupied with, if that is what we are accustomed to doing. We have to understand that any such "way of looking at things", in terms of neuroscience, is a matter of patterns of comprehension that have been established through neuroplasticity.

Such patterns of comprehension, because of the nature of neuroplasticity, constitute a kind of cognitive pareidolia, or cognitive bias, that academic thinkers may not even necessarily notice. It's not that a "set" approach is "wrong" in any way. It is a question of where the limits lay, in any mode of comprehension.

We bear in mind that whenever we conceive anything, we are doing so by introducing objects and structures. We don't necessarily have to understand the nature of any object in any profound way, but we must be clear about what structures we are talking about. We are used to thinking of "structures" in science and mathematics, as objects in nature, or in some "mathematical reality", rather than in terms of structures of mind or comprehension. This is not the ultimate way of doing things, but a phase in our progress. It's not a phase that becomes redundant, but is a phase that has to take its place in what is beyond it, which concerns understanding the mind that is doing the understanding.

The fact is that ultimately, our comprehension takes place in the mind. It is a construct of brain function. Our comprehension consists of structures through which we understand what we are doing. Without understanding these structures themselves, then in science and technology we are like beings who are learning how to take Lego structures apart, and redesign with the parts, which is a really important thing to learn how to do. However, when it comes to applying this to the brain, and our own mind and experience of being, we are likely to turn ourselves into a well designed Lego robot. And in doing that, we miss our true potential.

The Set Object

When we talk about a "set", the thing we are talking about and calling a "set" is an object. The tendency is to imagine that it is something separate from the mind and mode of comprehension that we are using, because we are considering it as an object. In fact, that alone is not sufficient to establish it as genuinely objective (as defined above). It is either way an object in the mind. DNA *as we understand it* is an object in the mind, that represents genuinely objective DNA. Nevertheless, our understanding of it and conception of it, is still a structure in the mind. This is completely related to the neuroscientific fact that what we perceive as the material world, is in fact a neural construct.

We have to use the mind to understand and comprehend the world. In science, we endeavour to bring that understanding and comprehension in line with what is genuinely objective, through objective scientific method. Not all pure mathematics at any time, is something that can at the time be verified to be genuinely objective. Rather, what we are dealing with is structures of comprehension that may or may not be genuinely objective, and may or may not be useful elsewhere in science, if not now, then in the future.

It is perfectly possible to have a structure of comprehension that is useful in practice, for describing and understanding objective material phenomena in science, but that only goes so far, in this respect. This is a very common and well-known situation in the history of science, and is often called a "scientific paradigm". In pure mathematics when we are immersed in a mode of comprehension, it is very easy to come to believe that all kinds of objects that we are thinking about also have genuinely objective counterparts. The idea of a "set" is one example.

A "set" is an object. But is it an object that exists independently of any individual mind that may conceive it? The fact that there may be large networks of minds who conceive it, such as the "mathematics community", and beyond, doesn't answer this question.

A "set" of apples is only a "set" because we have called it so. What actually exists, in the material, objective world, is the apples. If we choose to interpret that as a "set", that is an act of mind. An act of comprehension. It has a place, and we can do much with it. But our reliance on that also has a limit in its ability to bring us to a full comprehension of the nature of the world we live in, and our place in it.

When we write down some symbol and say it stands for "the set of all natural numbers", then we are combining the object called "set" with the object called "all" and the object called "natural numbers". Mathematicians do this so fast that they probably don't see the processing of the concepts in the mind that is going on.

This isn't a question merely of the structure of language. It is a question of how we conceive things. A question of the way we are thinking about things. Language in any case doesn't exist as a structure in its own right, independent of the human mind. It is generated through the human mind. We are not talking about the way in which we express thoughts through language, which is a different part of brain function, we are talking about the way in which we erect structures of comprehension in our mind, through thought processes.

The fact that those thought processes may be mathematical, and expressed using mathematical symbols, doesn't alter the fact that they are structures in the mind confined to particular part of brain function. "The set of all natural numbers" is an object in the mind that we may or may not think has a counterpart that is an object independent of any individual mind that conceives it. Either way, when we conceive it, it is an object. Being an object, it is also a structure. That's the nature of anything that arises through brain function.

The "set" object at its highest level of abstraction is what set theory calls the "empty set". This is a perfect demonstration of its nature as an object. The object "set" can be instantiated without a second object that would be its contents. This is then the "empty set". But if we instantiate it together with a second object, and link the two with the *relation* that in words we write as "of", in a phrase such as "the set of all...", then we then have a set *of* something, rather than the "empty set" object:

We can think of "the set of all natural numbers" as a structure of the three objects. The edges of the network are the relations between the objects. Even without knowing enough about those relations to be able to explicate precisely what they are, we can still see there are relations. So we could have, for {A} for example:

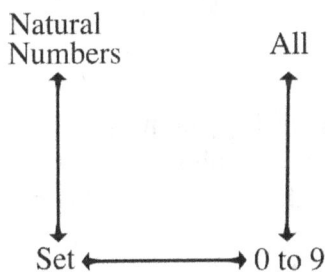

Natural Numbers   All

Set ⟷ 0 to 9

Here we are saying that {A} is the set of *all* natural numbers from 0 to 9 (here we regard zero as a natural number).

But when we say for {B}:

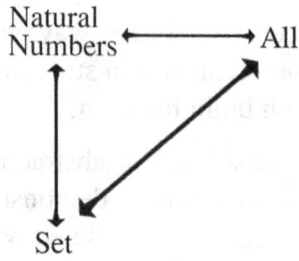

Natural Numbers ←→ All
Set

we are now saying that {B} is the set of all natural numbers. On what grounds do we assume that the relations between the three objects here, is the same as those between the previous four objects?

The concept "all" is an object. It is perfectly possible to create the structure "all natural numbers", and then proceed to handle that in whatever ways we like, as an object. But is the structure to begin with, anything more than a chimera - an object in the mind that has no genuinely objective counterpart?

Process

We cannot find, anywhere in genuinely objective nature or material phenomena, any number symbol, such as a decimal "5", or a hexadecimal "FFFFF", other than where this has been created through human intelligence. If this is what we mean by a "number", then numbers don't exist anywhere in genuinely objective nature or material phenomena other than in what we have invented. However, quantities and magnitude, and so on, do exist genuinely objectively.

This meaning of "numbers" begins with the process of explicit counting, as distinct from "number sense" as it is now often called. The latter, that can be observed in some animals, is something we humans are calling "number" sense, because we already associate specific quantities with the symbols that we also call "numbers". More strictly speaking, some animals demonstrate a "quantity sense" or even a "specific quantity sense". But that doesn't mean that they are attaching "labels" to any quantities, as we do, such as the label "5" or "FFFFF".

The symbols themselves, that we use, are always the result of a *process*. As we described earlier, there is some iteration process that "outputs" a resultant, that is number symbols. This is essentially the same iteration as *counting*. As stated earlier, the suggestion that you need *bijection* (one-to-one correspondence) between the number symbols you are counting with, and any number symbols that you wish to count, is simply wrong. The sleight of mind that is carried out in the argument that you *do*, relies on confusing the fact that you can *find bijection* between certain sets of numbers, with the idea that you *need* bijection, in order to be able to count. It is literally a piece of mentalism. It is *not the case* that with natural numbers you can *only* count natural numbers or multiples of natural numbers, or rational fractions of natural numbers, and so on.

In fact, the problem with "counting" the real numbers, has nothing to do with bijection, but rather, reduces to the problem of how you count the number of points on a continuum. And really, that problem itself, is a chimera. It's not a real problem, but one that has been made up. Having been up, and widely accepted, by large numbers of minds in a network, it is also widely but falsely believed to be genuinely objective (remember that genuinely objective means that it is not dependent on the workings of any individual mind or brain). It is essentially the same thing as fashion, or what is currently popular.

It has been made up in the same way that, for example, the concept of "an infinite number of points" around the circumference of a circle, has been made up. It's just a way of mentalising, or mentally modelling. You can still do lots of useful things with mental models, but that fact doesn't mean that a mental model is anything other than a mental model. It may be subscribed to buy a network of a very large number of subscribers, but that doesn't make it genuinely objective, any more than say, a religious belief.

The reason there is disagreement amongst mathematicians when it comes to infinity, and counting, and so on, is because different mathematicians are using different structures, or mental models, in order to do the comprehension, and also because when words are used, they often have more than one meaning, depending on the structural context in which they appear, or the mental model that they are being related to. Nevertheless, we can suspect that many mathematicians believe that what they are dealing with is something that is genuinely objective.

If we are clear about the difference between a number symbol, and the quantity that it represents, then that is one step forward. The number symbol is one object, and the quantity that it represents is another. They are two distinct objects. But it is perfectly possible to use structures of comprehension in which that distinction is lost.

If we truly appreciate that *number symbols* or the "labels" that in counting systems we give to quantities, are generated by an iterative process, an iterative process *that is distinct* from the quantities that they represent, and distinct from any iterative process through which quantities come into existence, then we are on a better footing. We can then realise that what we call "natural numbers" is an object in our intelligence, arising from our realisation in the intelligence we are being, of the principle of counting, applied to natural quantities.

8: The Natural Number Structure

E
verything we know about the relations between numbers (excluding the Cantorian so-called "transfinite numbers" which are chimeras), and the mathematical structures that they form, is about relations between quantities.

These quantities are quantities of distinct objects. We don't even have to know what kind of thing these objects are. They don't have to be material objects, or even genuinely objective objects. They are literally objects of the kind that Object Theory deals with. The *relations* between them are what in Object Theory are the edges in the networks of these objects. Thus, what characterises numbers and their relations, is the nature of the *structure* of an infinity of distinct objects, each object being completely abstract and having no other properties to consider.

Ordinarily, we might find node objects in a graph or network represented in two spatial dimensions. If the nodes of this graph or network are just abstract, we could just represent them as, say, points. Abstract objects, with no other properties to consider, are identical. But they are not one, because they occur at different positions in the network. In the case of a graph this is not a matter of positions in some other framework or system of coordinates such as the two-dimensional sheet of paper on which they may be drawn. The "position" of a node in the network is entirely a matter of the graphical topology of the network. It is a matter of the relations between the nodes.

What makes two abstract objects distinct, even though they are identical, is the *relation* between them. Without these relations, all that we are doing is representing the same one object, more than once. Thus, two objects in Object Theory are actually a "falsehood" or chimera, if the relation between them is null. Something that makes two objects distinct, is a relation between the two objects. If they are otherwise identical, but are distinct because they occupy different positions in some metric or space, then that is a relation between them. The metric or space is also, if you want to regard it as such, an object in its own right, to which the objects in it, have a relation. If you are just talking about numbers, you could say they occupy a position in "number space", but really, this "number space" is not separate from the numbers themselves, but is just the *structure* that consists of the numbers, and the relations between them.

So what we are seeing as the relations between the natural numbers, is a *structure* (in the sense of Object Theory). We are able to perceive and represent it as the relations between what we call the "natural numbers". The structure itself is actually the structure that exists genuinely objectively, in the infinity of distinct objects.

The Infinity of Distinct Objects

Where, or what, is the infinity of distinct objects? It is in the "mechanism" of how our mind works. There is no limit to the number of objects we can conceive as separate from the mind conceiving them. Is there a limit to the number of objects that are genuinely objective? To be genuinely objective, they must not be dependent on any mind that is conceiving them, or on any brain function through which that conception occurs. The question of whether there is a limit to the number of objects that are genuinely objective, is one way of asking whether the universe is finite or infinite. A question that formally, in science, there isn't an answer to.

So scientists and mathematicians go about their daily work, generally agreeing that collectively we do not know whether the universe is finite or infinite, and at the same time working with conceptions of infinity. Even the simple function $y = 1/x$ encapsulates this conception where $x = 0$. And that is because in mathematics we are accepting that there can be an infinity within something that is finite. We do this in our conception of the continuum. We say things like "there is an infinity of irrational numbers between any two rational numbers". Which is a conception. And we don't know whether it is genuinely objective or not. Because the only way in which we can demonstrate that something is genuinely objective, is through hard science demonstrating that it exists in the form of measurable material phenomena. Phenomena that does not depend on anyone's individual mind, or brain function. Even though we can only know about it, through the mind, through brain function.

We take the pure mathematics, some of which is established, actually, on the basis of network consensus, in the "mathematics community", rather than the demonstration of genuine objectivity, and we apply some of it to the physics. In general, the way in which the pure mathematics is able to describe the physics, becomes compromised, or breaks down, when we are dealing with infinities. Now, infinities in pure mathematics, in the form of singularities, have become part of the physics itself, in the case of "black holes". But elsewhere, it's still a problem. Trying to combine relativity theory with quantum theory, creates mathematical infinities that cannot be interpreted. At the current time we don't know what ŵthis means.

At the current time, there is the general idea that there is probably something wrong with the physics theories, in that they are missing something. Even though they are successful, there is something missing. Seldom do you hear that there might be something wrong with, or something missing from, the mathematics. Often, when dealing with cutting edge concepts, physicists themselves are using mathematics in a way that may comply with what is "established" in mathematics, and sometimes in new ways, where the acceptance of the mathematics is made with no understanding of the issue of whether or not it is genuinely objective. Or whether it is just a matter of network consensus.

These issues are not merely academic. They are going to be critical in our understanding as science continues to penetrate into the question of the human brain. And therefore, into the question of the human mind, human intelligence itself, and our human experience of being.

What is missing from physics, is not something just missing from physics. It is our lack of understanding of the mind that is doing the understanding, in physics. At the current time we have to carry on developing an understanding of genuinely objective material phenomena, whilst completely, as it were, to use a turn of phrase, leaving the mind that is doing the understanding, the brain function, out of the equation. And we will be increasingly assisted in this, with the use of AI.

What I am saying is that there *is* an infinity of distinct objects. It exists in the way our mind, the intelligence we are being, works. It is an infinity within the mind. But it is only within the mind. And AI, as it increases in power, specifically, in the use of quantum computing, and future photonics in general, will increasingly provide answers to problems "without going there".

The Kernel

At the heart of what we are saying is something that we cannot come to or realise while we are still positioning ourselves in the assumption that in mathematics and physics we are dealing with things that we can come to a full understanding of, whilst leaving the nature of our mind out of the picture.

One route is to realise that what is going on in mathematics is not inherently genuinely objective just because it is mathematically valid. While we cling to the idea that "mathematical validity" is the same thing as being genuinely objective, even subtly or tacitly, then the limits of mathematics are where we currently find them to be. Which is that, as things stand, the best we can do is to show whether something in mathematics is consistent according to the axioms of the mathematical model being used. To get beyond this, we have to realise that all this is a construct a brain function. It all takes place in the mind that we have, and are being.

We have to realise that what matters, for science including mathematics, is what is genuinely objective. And we have to realise that this means it is not dependent on the workings of any individual mind or brain. Or, for that matter, on any subnetwork of individual minds or brains, cooperating as a whole. But it *does not mean* that it is independent of brain function.

From this emerges the following: There is a multiplicity of distinct objects, such as material objects, and mathematical objects such as a circle, that are genuinely objective. However, mathematics also has to deal with infinities. Part of that, is the conception of an infinity of distinct objects. If you conceive "the set of all natural numbers", then you are conceiving an infinity of distinct objects.

An infinity of distinct objects is within the structure of our mind, but cannot be shown to be genuinely objective. Rather, it is an artefact of the mind we are using, that *is dependent on this individual mind*, or on a whole subnetwork of individual minds in which there is a consensus on this.

What we conceive as the natural numbers, in Object Theory, is a *structure*. It is an infinity of distinct objects, and the highest abstraction of that, is not "natural numbers", but simply the infinity of distinct objects.

An infinity of distinct objects is itself an object. The current "neuroplastic conditioning" of the *mainstream subnetwork of* minds dealing with this object, tackles questions concerning infinities of distinct objects, in terms of the concept of the "set". A "set" is taken to be genuinely objective, but the question of whether or not it is genuinely objective, is not considered in set theory.

The notion that something is genuinely objective (by the definition of Object Theory) simply because (a) we are considering it objectively, and (b) because as a tool *we find it works*, is mistaken. If it is *not* genuinely objective then it may fail to work when we move the boundaries within which we are working, in some way.

An example is the Ptolemaic universe. It is not genuinely objective today, but once, it was. However, we would currently, in the mainstream, believe now that it was *ever* genuinely objective, but rather, was always just a matter of consensus. This is because our current manner of understanding is still entrenched in *naive realism* - the belief that we are coming to understand something that is other than a construct of the brain function through which our intelligence arises.

Each time we revise a scientific or mathematical paradigm by discovering something more that embraces the old, but goes beyond it, then it is because we *were* using a *modus operandi* of understanding that worked in a limited domain, but is not now genuinely objective.

To complicate things further, our understanding *is* genuinely objective, for now, if we can empirically verify it. However, what empirical results we are capable of, changes as science changes and evolves. So what is genuinely objective now, may not be so in the future.

This situation arises because of the idea the world we perceive and conceive - whether we are talking about the material world or the world of the intellect - is other than a construct of brain function, just because we conceive it as such. That position is the position rooted in *naive realism* - the presumption that what we know about, experience, or conceive, that is other than a construct of brain function, and therefore other than part of the principle by which our mind and experience of being arises through it. To come to know that principle, we must find not only what is genuinely objective, but what is genuinely not a construct of brain function.

When it comes to the consideration of infinities in mathematics, we are going to be talking about the "process" object as a more fruitful object with which to proceed, if we wish to enter a post naive realism mathematics, than the object we call a "set".

So what we are talking about here, is about the kind of mathematics that we need in order to literally develop genuinely objective scientific knowledge of the mind, through scientific knowledge of brain function. Many people think that we can some to some kind of scientific knowledge of what is genuinely objective, through the mind, without first scientifically understanding the mind. Essentially, we are saying here that this is also *naive realism*.

The *Process* Object

In Object Theory we can define a process as a directional structure with an input and an output, where the output is distinct from the input. And an iteration process is then an object with an input and an output, where some part of, or all of, the output, feeds back into the input. In mathematics this kind of structure is already familiar in the concept of an iteration function.

There is nothing in this definition that requires the output or the input to consist of discrete objects. There is also nothing to exclude that. So in mathematics, a "smooth" function, in terms of Object Theory is a process. But so also, is the kind of iteration function used to create, for example, the famous Mandelbrot set, whose output consists of discrete objects.

We may well then think that there is a question concerning whether or not there is any kind of process "in between". This is not unrelated to the continuum hypothesis in existing mathematics. There, it appears as a question of whether there is a set whose cardinality is strictly between that of the continuum, and that of the natural numbers.

In fact, Object Theory addresses the same essential question, but is coming at it from a "different angle", rather than via the concept of a "set". The problem with coming at it through the concept of a "set" is that the problem gets stuck on the concept of "cardinality". We end up asking is there a "set" whose cardinality is "in between" that of the natural numbers, and that of the real numbers? By implication, the result of the work by Malliaris and Shelah, showing that the infinity of the reals is the same "size" as the infinity of the naturals, means that the question is a false question. It assumes a difference in cardinality between the reals and the naturals, that isn't there.

An Object Theory approach to this situation, would involve questioning whether the concept of "cardinality" itself, is in fact genuinely objective.

Discussion

We are saying that although the infinity of objects is objective (by definition), because we conceive it that way, it cannot be shown to be *genuinely objective*, meaning specifically that it cannot be shown to be independent of any mind or network of minds that conceive it. We could argue that we observe the effects of genuinely objective infinities in black holes, but that is not an infinity of distinct objects. On the contrary, the mathematics of it relates to the continuum. Only by further contrivances such as the concept of "an infinite number of points", that then "appear on the continuum", can it appear to relate to an infinity of distinct objects.

In contrast, for example, objects such as "quantity" or "hyper-large number", or a quantity of material objects such as stars, and so on, are genuinely objective, because we can measure them, and demonstrate them in a way that is not dependent on any individual mind or network of minds. In the same way, a chemical reaction is genuinely objective, and the relation of the hypotenuse to the other two sides of a triangle on a Euclidean plane are all objective, because none are dependent on any individual mind or network of minds that conceives them.

The infinity of the naturals does not embody irrational quantities in its structure. However, the *continuum* object does. There are irrational *quantities* that occur as relations between quantities that we call *ratios*. To these we then *apply* numbers, number-making processes, or number-based measuring. All of these numbering systems are derived from natural numbers, which we derive from the counting process, and relate to the object we call the infinity of the natural numbers, which is an object we invent in our mind that is based on the conceived infinity of distinct objects.

Because the infinity of objects cannot be shown to be genuinely objective, then the *infinity of* natural numbers also cannot be shown to be genuinely objective. It is worth reiterating here that this in no way means that *finite quantities* of distinct objects, that we call "numbers" and apply number symbols to, are not genuinely objective.

The discrete *quantities* that number symbols stand for, the quantities themselves, are indeed, genuinely objective. Even the number symbols, themselves, that we may invent and use, as visual objects, are genuinely objective. Even the patterns of relation that occur between them, in a numbering system, are genuinely objective. But the *meaning* of number symbols, what each symbol "stands for" - notwithstanding that the symbols are *related* to each other in a way that is genuinely objective, because it maps to quantities and their relations - requires an individual mind or network of minds, to interpret the symbols. Hence, the meaning of any individual symbol in isolation, is not genuinely objective.

Inherent in the concept of "the set of all natural numbers", is our conception of the infinity of distinct objects. The infinity of distinct objects inherently contains all possible relations of quantities, except those relations of quantities that we call irrational. The only way that relations between "natural number" objects - which are derived from the infinity of distinct objects - can represent an irrational quantity, is through an *infinite process*, the process, for example, of determining an irrational number expressed using a string of natural number objects such as 3.14159.... *etc*.

Finite quantities are genuinely objective. Some animals, as studies have shown, have a sense of quantity, although this is often reported to be a "sense of numbers". Human beings manifestly do have this sense of quantity. But that is not the same thing as the conception of a "number". Children have to be *taught* "numbers" as a mental and intellectual exercise. They are a concept in the mind that is passed on from generation to generation, and that is able to be so, through the principle of neuroplasticity, rather than through purely Darwinian processes.

Nevertheless, the infinity of distinct objects cannot be shown to be genuinely objective. And it follows that the infinity of the natural numbers cannot be shown to be genuinely objective. It is indeed, *objective*. But it is an object in the sense of Object Theory, that is, it is an object in the mind, that is an object because it is treated as such, meaning, it is treated as being separate from, and independent of, the mind that is conceiving it.

Many of the "difficulties" and contentions amongst mathematicians, regarding numbers in mathematics, then come down to matters concerning the way the intelligence that we are using in mathematics, is working. They arise from working with matters that are not genuinely objective. We are in fact, exploring the way our mind works, and the way *processes* of reasoning that we are using, work.

We know that irrational quantities exist genuinely objectively. That is, they are not dependent on any individual mind or networks of minds that may be conceiving them. For example the nature of circles, with the ratio of circumference to diameter, *is what it is*, and does not depend on any individual mind, or on any consensus in a network of minds. The ratio of the circumference to the diameter of a circle, irrespective of the fact that it is an irrational quantity, and can never be perfectly measured, is not a bone of contention.

The whole idea of ratios, and the fact that we call irrational quantities "irrational", is based on the conception of numbers and *number ratios*. When we confuse that intellectual device with *relations* of natural quantities, such as the lengths of the circumference and the diameter, and we let the smoke of a genie out of the bottle of the mind. And it obscures our vision of what is really going on in nature, and indeed, in the nature of our mind, and intellect, which arises through nature.

What is really going on in nature, and indeed in our mind and intellectual processes concerning reasoning, is objects and structures. That is, objects that are genuinely objective, and distinct, and then structures consisting of the relations between those objects. And then out of this, in our mind, arises structures that are not genuinely objective, even though we can think with them objectively, and even sometimes use them as a tool for describing things that are genuinely objective. It is this using mathematics as a tool in this way, and mistaking that for its genuine objectivity, that leads to all kinds of miscomprehension and misunderstanding of what it is that we are looking at, even in in certain areas of physics.

Let us take the simple example of the circle again. We can view the circle in terms of two objects, its circumference and its diameter. There is a *relation* between these two objects, that we call a *ratio* of length. This is a ratio of quantities. It is a ratio between two distinct objects. However, it is not a relation between two objects that are not already structures in their own right.

It is not a *relation* between two of the objects that are structures, networks, or what we call "sets", within the infinity of distinct objects. Explicitly, it is not, as we know, a ratio between any *set* of objects (that we label as *natural numbers*) in the infinity of distinct objects, and any other *set* in that infinity. The *notion* that it is, is however, something that *does exist* in the realm of the way brain function translates to out mind and reasoning processes.

So just as the *natural numbers* and their derivatives that we conceive, *are* derived from the *relations* between structures within the infinity of distinct objects, so this is *not the case* for the quantities we *call* "irrational numbers", and treat *as though* they are inherent in the infinity of distinct objects. Hence we cannot express an irrational quantity using natural numbers or strings of natural numbers, *except as an infinite (unhalting) process*. Rather, the *relations* that we imperfectly express using natural numbers or real numbers, are indeed *relations* between quantity objects, and these relations are themselves objects (the fifth premise of Object Theory) that are quantity objects. However they are not ones that are not found in the infinity of distinct objects.

It is simply not the case that a quantity object *must be* a number object, *if* by "number" object we mean an object that is other than a process object. However, if we understand "numbers" to be processes, or partial expressions of processes, then indeed numbers are everywhere, and the basis of everything that exists for us to know about, through brain function.

The Naturals, the Reals and the Irrationals

The natural numbers come from our system of counting the infinity of distinct objects. The system we choose does not have to be on the decimal number base. We could use, for example, binary, unary, or the infinite number base.

In the unary system we can still express ratios, but we cannot simplify them. As in decimal, or binary, and other bases, we cannot perfectly express irrationals. The reason is that we are still using a counting system (that is an *iteration process*) based on the infinity of distinct objects.

Even the infinite number base we still cannot express irrational because we are still using a counting system that amounts to labelling the infinity of distinct objects. It is just that in the infinite number base we never use two symbols that are alike, and our iteration process now simply creates a unique symbol for each natural number, on each iteration, that is distinct from all other symbols already generated in the process.

No number base of this kind can perfectly express irrationals. Each number base creates unique labels for the infinity of distinct objects, and is an *iteration process*. It is an iteration process that halts for naturals and reals. The quantity objects of naturals and reals are inherent as *relations* within the structure that is the infinity of distinct objects. However, the process does not halt when the *structure* as a whole is considered as an *object*. Similarly, the iteration process used for constructing real numbers does not halt when applied to irrational quantities, because these are relations that do not even exist in the structure that is the infinity of distinct objects.

We are looking at relations between quantity objects that do not all belong to the same structure. Irrational numbers arise from applying counting systems based on the infinity of distinct objects, to relations between quantity objects that are not inherent in the infinity of distinct objects. The truth of the matter lies in the relation between the *quantity* object, the *infinity of distinct objects*, and the object that is the *infinite continuum*.

9: Invariance and the *Genuinely Objective* - a Discussion

We can talk about the "genuinely objective" with reference to the brain, rather than the mind. What is genuinely objective - as far as concerns brain function is concerned, is what is *invariant* with respect to the functioning of any particular brain or network of brains, in the way the brain creates our experience of the world, mind, self, and thought. Then, through the functioning of our particular brain, we experience the thing we call "the world" and other areas of mind, such as mathematical objects, in our own way.

If a mathematical object is genuinely objective, then it must be invariant with respect to the particular mind and brain that conceives it. Even if that particular mind and brain still conceives it "in its own way". This principle applies to networks of minds or brains, too.

Mathematical objects such as the one we call Pi, for example, are invariant with respect to brain function and the mind that it gives rise to. But that doesn't mean that Pi has to be conceived as 3.14159... using decimal numbers, for example, because there are other ways of conceiving it. In fact, even without knowing anything about it, we perceive it, or a rendering of it, when we so much as see a circle.

The mainstream way of approaching this at the moment, is to presume the circle *as we see it*, or as we conceive it, mathematically, as some kind of prior "reality", that the brain function then creates a "representation of". In the post naive realism view, everything that we think is a property of the circle, or in the nature of the circle, is in fact what is arising in the nature of brain function.

In this way, all our understanding of mathematical structures, needs to be understood in terms of brain function. And brain function itself is essentially, a *structure*, in terms of Object Theory.

It may be that we feel comfortably confident that most of mathematics is genuinely objective. However, we know this is not the case, for example, with the Cantorian theory of the infinities of real numbers, and natural numbers. It cannot be genuinely objective or it would not have been possible to demonstrate that the infinities of the reals and the naturals are in fact the same "size", as has now been demonstrated.

The whole effort to try to find a mathematical "proof" for something, is the effort to try to demonstrate that it is genuinely objective, and hence, invariant. However, what we find, or have now found, is that any mathematical "proof" is only "proof" according to the axioms of the mathematical model used to arrive at the "proof".

So the really interesting question is about the axioms. Are axioms genuinely objective? There are some axioms from which we build mathematical structures, for which the mathematical structures can also be empirically verified in the material phenomena of the world. We could, for example, literally measure the length of a hypotenuse, to see if it is the square of the other two sides. One thing we do know is that the material phenomena of the world, the nature of it, is genuinely objective.

When we are looking to verify a theory in physics, say, empirically, then if we find the verification we are looking for, we take the mathematical structures to be correct. However, we have currently done this in both the theory of relativity and quantum mechanics, for example, and arrived at a situation in which both are correct, but they are incompatible with each other. The same kind of thing happened when we used to have a particle theory of light, versus a wave theory of light. Both seemed to be correct, but they were incompatible with each other.

Now in the case of the theory of relativity and quantum theory, we are looking for something that is invariant with respect to whichever theory is used to partially describe it. In the case of the wave-particle duality of light, that invariant thing was found in the form of the new quantum theory. In today's case, physicists speak of a "theory of everything".

Such a theory, we might expect, should describe something that is genuinely objective. Not just something that is a consensus in a network of scientific minds. But remember that what we are calling *genuinely objective*, as something we have an understanding of in our mind, and hence through our brain function, we have already defined as something that is invariant with respect to the mind or brain function that conceives it.

This is not at all the same thing as saying that it is something that exists separately from brain function. Everything that we conceive and understand, all our thought structures, all our comprehension, is all a construct of brain function. Rather than presuming there is something separate from brain function that we should be considering, brain function itself, which is already here, and which we are already presented with, holds the key.

There is no doubt in neuroscience that what we experience as the objective, material world, is in fact a construct of brain function. The aspect of the functioning of all brains that is genuinely objective, that is, whilst it is implemented through the functioning of any individual brain, but is not dependent on that brain's individual functioning, is what we experience and conceive as the objective world.

Hence, genuinely objective mathematical structures should be considered as equivalent to the mathematical structures that are universal to all brain function. Specifically, genuinely objective mathematical structures that are conceived by human beings, must be equivalent to genuinely objective structures that individual human brain function is capable of manifesting. Whether or not it does, in any particular brain.

The most salient part of the nature of human brain function, is network complexity. Network complexity is the same thing that Object Theory regards as a complex structure.

A potential way forward in looking at our situation in this way, is to understand the principles of hyper-complex structures, or networks, or graphs, that are invariant, with respect to specific instances of hyper-complex structures. Specifically, each individual human brain is a specific instance of an invariant principle of the brain. Whilst each of us has our own subjective experience and understanding of the objective world, the objective world itself, that we experience, is a consequence of this invariant principle.

At the current time, the mainstream "understanding" of our situation is that what we experience as the objective world, which we know is a construct a brain function, is some kind of "model" of a world that that weekend never know directly, because it is separate from brain function. This is naive, and introduces an unnecessary additional "objective world", which in fact, is just a supposition. It is a hangover from our subjective, personal experience of the world as being separate from our self, where of course our self experience is also a conditional experience of being that arises as a construct of the same brain functioning.

The meaning of all this, in terms of our experience of being, is another matter, outside the sphere of what we are currently discussing, which is our knowledge of the nature of mathematics, and the nature of the objective world we inhabit. It is this latter knowledge through which so much practical compassion manifests for the human race, through technologies. It is also the knowledge through which we are potentially capable of building a new world beyond the problems and suffering associated with the world as we currently know it.

This necessary new knowledge of structures and complexity is not something that we can tackle just by using the mind that is arising through our own hypercomplex structure called the brain, without assistance. Machine intelligence or artificial intelligence can take some basic principles of the way our evolutionary intelligence works - based on the neural network - and create out of it abilities that we ourselves are not capable of. And so machine intelligence can learn from vast quantities of data, things that we ourselves would not be able to learn, in order to come to knowledge of the inherent "meaningful patterns" that are there. This is the way that we are currently using machine intelligence.

So the knowledge of hypercomplex graphs and networks in general, is something that we can benefit from. It is something that we need to come to knowledge of, before we try tackling the specific hypercomplex network of the human brain. If we are trying to do the latter, before we have some graphs of the former, then we are trying to run before we can walk.

Our current electronic supercomputers are not sufficient to tackle what we are talking about. At present, our studying the nature of relations between important properties of very large graphs is beyond the capacity of current supercomputers, because the sizes of the numbers associated with those properties "explodes" so fast, with increasing graph size. It is knowledge that we are more likely to come to, as a network, after we have well established photonic quantum computing, and its corresponding AI. Then we are more likely to be able to effectively explore and determine and understand the invariant principles behind the non-linear behaviour of very large graphs (hypercomplex networks), and relate that to human brain function.

It is a mistake to think that we can simply put aside the need to understand ourselves, and just use AI to interpret data, and give us some answers. This, currently, is largely the way we are using AI or machine intelligence. It is not going to be sufficient except for certain aspects of our understanding of the way the brain works. The deeper knowledge that we really need, requires that we change and enlighten our own ways of understanding. What comes through the use of AI is intrinsically, in itself, already about mathematical structures, in the state of having been empirically verified. But actually, no matter how surprising or amazing such revelations may be, it still requires *our* intelligence, to put it into human context, and see the meaning in it.

10: The Continuum

A function such as $y = 1/x$ is not usually considered as a process, however, we can considerate in this way. The objects it consists of are y, 1, the division operation, and x, and we can consider the "=" object as a relation. The division operation is itself a structure and is often literally carried out as a process.

Such a structure as algebraically represented is often also represented as an object that has an input x and an output y. Such an object is a process object. Let us say that the domain of x is the (positive) real numbers. Now we would normally say that "as x approaches zero, y approaches infinity". But let's not talk about the approach. Let's talk about what happens when x *is* zero.

We might talk about y being an "infinitely large number". But there, we are just being colloquial. We can say "y is infinity", but what does that mean? On the approach, or elsewhere in the domain, we are talking about numbers. In that part of the domain, when x is not zero, y is always a number. But infinity is not a number.

We might want to say that y is an "infinitely large quantity". But is such a thing genuinely objective? There is no way to verify that the infinity of distinct objects, for example, is genuinely objective. And so there is no way we can verify that the object we are calling "an infinitely large quantity", is genuinely objective. It is only something we might speculate about, or think about. And then, without necessarily being where of the mind processes we are using, and the way they are working,

Another way we can interpret this so-called *singularity* is that when x is zero, the output of the function object, y, *escapes* to infinity. That's a term that may be more familiar in the study of fractals, where it means something a little more specific, but that is nonetheless related. The "escape" here means, we could say, that the output escapes from the condition of being an object that is a "number". Alternatively, in terms of sets, we could say it escapes from a condition in which it is a member of the set of all real numbers.

Considered in terms of processes we have the following:

Natural numbers are generated through the infinite iteration process that creates natural numbers. We can talk about the whole output of this process as one object, which is what we do when we speak of "the set of all natural numbers". As a process the natural numbers are the resultant of the infinite iteration process of labelling the infinity of distinct objects. The natural numbers express discrete quantities of distinct objects.

This equation however, constitutes a "smooth function". It applies not just to the natural numbers, but to (let us say) all real numbers. The real numbers are used to express quantities on the continuum. The "continuum" is another object in the mind to which we apply mental processes. For example, we may just assume we know what it is, and what it means, or we may try to build definitions. And then we may try to build structures that include objects such as "the real numbers" and "the continuum".

One way or another, what we think about the relation between the object we call "the real numbers" and the object we call "the continuum", will depend on how we are thinking about each of these objects. It will come down, literally, to the way we think. It will come down to the mind processes, and structures of thought and thinking, that are going on, as *processes*. The general endeavour is to establish some "proof" or at least some demonstration or conjecture, about something that is not only object oriented, but is also assumed to be a truth or fact in its own right, independent of the mind, or mind networks, that are contemplating it.

The position of neuroscience is that this is impossible. The activity can be said to be the activity of a mind or network of minds endeavouring to establish something believed to be other than just a demonstration of the way those minds have come to be working, through the principle of neuroplasticity. The alternative is that we understand that what is going on remains in the domain of the mind, even when it relates to material phenomena and the way material phenomena behaves. This is entirely consistent with the fact that even

Page 65

what we experience as material phenomena is a construct of brain function.

Numbers as Iteration Processes

We know that there are infinite iteration processes (IIPs) that "output" a finite object, for example, a convergent infinite series. We also know that there are IIPs that "output" infinities. These infinities are clearly of different kinds, independently of Cantorian theory. The nature of the infinity depends on the nature of the IIP. For example, an IIP that is a simple divergent series results in an infinity that is clearly of a different kind to that produced by the IIP that results in the Mandelbrot set. The infinite isolation processes involved are different in each case. And the objects they produce are different. This is the case whether or not we attempt to use Cantorian theory to try to understand them.

The output of the divergent series is an infinity. We might *call* it an "infinitely large number", but that would be incorrect if by "number" we are stating the name of the kind of object that is an output of a halted iteration process that produces a real number, or the output of the iteration process for producing natural numbers, that has halted.

It would be incorrect, because the output of a divergent series is the output of an IIP that has *not* halted. We cannot equate or align or create a bijection between, the output of an iteration process that has halted, and that of an unhalted infinite iteration process. In the case of the latter, the output is itself a process.

So in dealing with structures of relations between numbers in general, we are dealing with structures of relations between iteration processes. These iteration processes may or may not be infinite, that may or may not halt.

In the case of the natural numbers, there is an infinite iteration process for producing them, and every natural number is an output of the process halted. The output of the unhalted process is itself an infinite iteration process.

The rational numbers other than the natural numbers are the result of an iteration process that may or may not halt, that is the process of dividing one natural number by another.

The *irrational* numbers are *processes* that are the results of the unhalting IIP that produces them.

Each *natural* number is an output of *part* of the IIP for the production of the natural numbers - it is the part that has halted.

Each *rational* number or terminating decimal that is the output of an iteration process that halts, can also be considered as part of an unhalting IIP. Numerically, we can see this in the fact that in the decimal number, for example, the iteration process that produces the digits after the decimal point, and comes to a halt, could also be extended infinitely in the production of zeros. Alternatively, the number could take its alternative representation ending in infinite 9s.

The same principles will apply to the imaginary numbers.

Hence all numbers can be considered as either IIPs or as parts thereof.

We will therefore later be referring to numbers in terms of IIPs, using the notations:

$$\circlearrowleft (N)$$

for the natural numbers, and

$$\circlearrowleft (R)$$

for individual real numbers, and will have more to say about this shortly.

An important point here is that when an IIP has not halted, that is, when it is infinite, then it remains a *process* object, and can be *considered as* a process object, and its "output" or "resultant" or "result" must correspondingly be considered as a *process*.

So the resultant of a divergent infinite series, is a process object. There is no bijection or correspondence between it and the output of a number creating process that has halted. It is a fundamentally different kind of object that must be handled in the appropriate way.

To further the illustration, the same is true of the resultant of the IIP that produces the Mandelbrot set. The Mandelbrot set is not itself a number, and is a fundamentally different kind of object to the number objects that it is a set *of*. In point of fact, the Mandelbrot set is a *process*. It is the result of an IIP, whose output is an IIP. The IIP that produces the set, is the well known iteration equation

$$z = z^2 + c$$

to which is attached the rule for deciding whether or not the result of each iteration belongs to the set. This rule of decision is part of the IIP. Also part of the IIP is change from one c value to another.

All infinities that are the resultant of an IIP that has not halted, are themselves infinite processes. However, some such resultants are processes that "match" a finite object that we do not ordinarily consider as a process, and may not be aware of as a process, such as a real number.

So when we have a convergent series S, for example, which is itself an IIP that converges to n, we say

$$S \to n$$

and in calculations may then simply say

$$S = n$$

We can do this, and make successful calculations, without ever really appreciating the kinds of objects that we are dealing with. The resultant number n may also be the resultant of another IIP, for example, if it is a natural number, it is also the resultant of the IIP for producing natural numbers (in whatever is our chosen number base).

To reiterate then: Structures of relations between numbers and number processes, can therefore be considered as structures of IIPs that may or may not halt. More informally, we can just say that all numbers are either IIPs or parts thereof.

Quantity Continua

Consider again the equation $y = 1/x$. If we want this equation to be applied over the whole domain of real numbers, then as a process object it is a process that arises from the fact that it is a structure that is relating what are *already processes*, namely, the process that creates all y values, the process that creates all x values, and the division process.

The processes for the y and x number values is the same in each case. They are the processes (plural) for producing the real numbers. That is, it is more than one process for each, x, and y. Later, we will see why there is no single IIP for creating "all" the natural numbers, even though this may seem superficially counterintuitive.

An IIP does not have to be a process that only outputs on each cycle. It can also be a process that continuously outputs, in the same way as a smooth iteration function. A simple example is the position of the centre of a moving wheel on a track, relative to the track.

In contrast, the IIP that produces the natural numbers by cyclically creating unique labels for objects in the infinity of distinct objects, outputs only on each iteration.

Infinite iteration in itself, or the IIP object, whether its output is continuous or discontinuous, is a *principle*. It is a principle that we already know exists widely in nature and in mathematics, and even just in elementary arithmetic. In Object Theory we can consider this principle as an object in its own right, before we relate it to any other objects. For a general infinite iteration process or IIP object we can write

$$\circlearrowleft (\infty)$$

and for an infinite iteration related to a specific continuum quantity object P we can write

$$\circlearrowleft (\infty) \{P\}$$

Where the quantity object may be either a continuum quantity, or a quantity of distinct objects.

The process for generating a real number, in the case of an irrational number will be an IIP, and in the case of a rational number will be a part thereof, in the manner we have already described above. However, as we shall later discuss, even a natural number can be considered as a real number IIP that produces infinite zeros after the point.

When we come to write *any* real number generating process we will use

$$\circlearrowright (R)$$

which may or may not be an infinite process depending on whether we want to define it as such, in the way just mentioned. For an irrational number it *will* be inescapably an infinite process. The reason for this notation without the infinity sign, is simply that it refers to the process for producing a single real number. This is a process that may or may not halt, and as we described above, even if it halts, it can be considered as a part of the IIP that would result in the infinity of zeros after the last digit, or that of the alternative form ending in infinite 9s. Therefore, we can say this notation is for an IIP or a part thereof, depending on the real number that it corresponds to.

We apply an IIP of this kind to a natural quantity continuum, when we measure it using a number. In doing so we make a structure in which the number is related to the continuum quantity. However, this does not mean that the continuum quantity is the same thing as the number. They are still two distinct objects. This remains the case even if the "stuff" or natural phenomena of which the continuum quantity consists, such as for example, energy, or length, has not been specified. In the most generalised case, even a proposed abstract continuum quantity is still a distinct object from the IIP or part thereof, that is applied to it, in order to express it as a number.

Philosophical Context

The question of the fundamental nature of the "stuff" of natural quantity continua as we know it, has hitherto been considered as a question about natural phenomena in a way completely isolated from the question concerning how brain function constructs our conscious experience of it, our knowing of it, and the intelligence through which we study and come to understand it.

As our knowledge of the brain has increased, a decoy in our understanding has persisted, in which our attention is focused on the idea of brain function creating an approximate "model" of the material world, and creating from its neural learning "predictions" about its nature. This deflects attention away from the stark fact that not only *how* we experience the material world, but also *what* we experience *as* the material world, together with all our understanding of it and thought about it, is itself, literally a construct of brain function.

There has been a kind of tacit, persistence of the old Cartesian assumption of a duality of mind and matter in the form of material brain function. A presumed duality that modern neuroscience and neurosurgery cannot support.

This is part of the question about how to make the transition between naive realism, in which appearances and experiences are taken at face value, as being the true nature of our world, and *post naive realism*, in which it is realised that the true nature of our world is not necessarily at all anything that we would naturally recognise as sentient beings. Modern physics, both classical and quantum, even when viewed from a position still attached to the old Cartesian duality, is already sufficient now to completely reject naive realism.

From the point of view of *post naive realism*, the question of the fundamental nature of natural phenomena, is already part of the question of the relation between conscious experience and brain function. This is a question that cannot be addressed simply by naively assuming that quantity phenomena as we experience it, such as space, motion, matter, and time, even at the local scale, have absolute existence in the way that we experience them.

The nature of this experience itself, is a construct of brain function, and we have no more reason to believe that what we experience as the material phenomena of our world is separate from constructs of brain function, than we have a reason to believe that we already know the way in which brain function amounts to our experience of being and world.

Page 71

As far as brain function is concerned, we don't *know* what the process is, relating brain function to the creation of our experience of any natural continuum quantity, such as space, time, or matter and its properties. What we *can* do, perfectly sensibly, given what we now know about the brain as the supreme complexus, is reject the presumption that a natural quantity continuum is simply a thing based on the processes through which *numbers* are produced. We can reasonably expect it to be *far more complex* than that.

Numbers are objects that occur in our intellect, and that we think with, in our mind, and that we then apply to the existence of distinct objects and continuum quantities, in our world. First, for this to happen, there has to be the creation of the mind, through brain function. It does not follow that because we can understand some of the behaviour of natural phenomena, in terms of numbers, that we can apply this kind of understanding to the question of how brain function becomes our mind, and experience of being and world.

As part of raising our comprehension into post naive realism, we need to go beyond confusing *numbers* with the natural quantities that we identify numbers with, in order to measure them. This applies whether we are talking about quantities of distinct objects, or continuum quantities.

Quantity Continua as Objects

Let us consider the difference between an iteration process that creates a continuum, and one that creates or is related to distinct objects, such as numbers.

Whilst an iterated process object P can be any iterated rules for the process, it does not have to be one in which for each iteration that we count as n, P_n must be uniquely related to a natural number. P could also, for example, be the position of a wheel on a track. What if the process P *is* just such a *quantity process* for a quantity Q such as a finite quantity of infinite track passed over by a wheel?

Then, simply, the infinite iteration is

$$\circlearrowleft (\infty) \{Q\}$$

where Q is now a *continuum quantity* of length (of track), rather than an otherwise unspecified *continuum object*. Consider again the object

$$\circlearrowleft (\infty)$$

which is the isolated infinite iteration object or IIP, with no quantity object related to it. Essentially, this is a symbol for a *principle*. This principle is itself an *object*. Because we have defined it as an infinite iteration process, then by definition it seems to imply *distinct cycles*. It seems then that we must have a *number* of distinct objects, called cycles, which then also it seems must amount to a continuum, when we apply the IIP to a continuum object.

However, the iteration object has no "size". We may apply any "size' to it from infinitely small to infinitely large. And *distinction* between cycles can only exist by introducing further *distinct objects*, which we can call "marks" that are then *related to* the iteration process. To create cycle objects, we must have at least two "marks" - one on the iterator, and one on a distinct object to which the iterator is related. There is no restriction on what other *kind* of object could constitute a "mark". Even the object "wheel" or "circle" or "coordinate system" will become a "mark", if it is part of the structure.

The natural numbers are an IIP in which there *are* cycles. For example:

$$\circlearrowleft (\infty) \{N\}$$

is the IIP applied to the infinity distant objects, through a process that labels the objects with the things called "natural numbers", in nested cycles. So here the $\{N\}$ object is not a continuum nor is it the "set of all natural numbers", but rather, N just conveniently stands for "the natural numbers", and the structure

$$\circlearrowleft (\infty) \{N\}$$

means the very well known and commonly used *iteration process* that we use for generating natural numbers, in whatever our chosen number base happens to be. The actuality of the numbers, before our labelling of them through this IIP, is the structure that is *the infinity of distinct objects*, and all its inherent relations.

So "marks" are just objects that the IIP principle takes place *in relation to*, always in such a way that we can then apply *iteration numbers* if we want to, because we have made *cycles*, or cycle objects, from our structure. In the case of

$$\circlearrowleft (\infty) \{N\}$$

which means the iteration process for generating natural numbers, the iteration numbers (iteration 1, iteration 2... and so on) are the same as the natural numbers being generated by the IIP.

In the wheel analogy, marks *could be* thought of literally as (1) a mark *on* the circumference of the wheel, and (2) a mark on the track. These "marks" are *distinct objects*. In conjunction with the IIP they make a structure that is the beginning of a *number generating* iteration process. Without the "marks" there are no distinct "cycle" objects. However, there is still the IIP as an *infinite iteration process* object, when there is another object to which it is applied, or related, *that does not constitute* a "mark". Such an object is a *continuum*.

An actual wheel, in material existence, exists already intrinsically, *with* or *as* a "mark", or "marks", because the material existence of anything already involves distinct objects to which that thing is related.

For this reason it may be difficult to imagine an IIP as a kind of ideal "wheel", without also accompanying what we are thinking of, with "marks". This is because *any further object* we imagine might also becomes a "mark". For example when we imagine a point on the wheel, that is a "mark". If we imagine a *physical* wheel, even idealised, its *physical necessities* are a "mark".

When we imagine an actual wheel we tend to imagine some object or objects to which the wheel is related, and relative to which it can be seen rotating. Even if that other object is only our own imaginary viewpoint. That imagined object, even our imagined visual viewpoint that we don't notice, is a "mark".

The object:

$$\circlearrowleft (\infty)$$

is the principle of the *infinite iteration process* without relation to *any other object*. It is only when it *is* related to another object that we can start to build structures of meaning and understanding. To relate it to another object we *could* write, for example:

$$\circlearrowleft (\infty) \longleftrightarrow \spadesuit$$

and then specify that the spade sign is some "stuff" or "thing" that appears as a continuum. But what then, is the nature of this relation? It is one object, that we could also have written without the arrows as: $\circlearrowleft (\infty) \{\spadesuit\}$. Essentially, we are saying that $\spadesuit$ represents the particular kind of continuum *stuff*, for example, length, or energy, and the *relation* shown by the arrows is one that says this stuff $\spadesuit$ is *instantiated* through the continuum IIP as *a continuum object* $\spadesuit$, which is a continuum of this particular "*stuff* of $\spadesuit$", such as length or energy.

Scale

If an IIP has not been related to any other objects of *measurement*, it is "scale-free". We may feel that its application and relation to a quantity continuum as:

$$\circlearrowleft (\infty) \{Q\}$$

must imply a unit of measurement, or some *numbers*, simply because $\{Q\}$ is a *quantity* object, albeit a continuum. However, this conviction, if we have it, arises not from what we encounter in nature as continuum quantities, but from *our* inherent pre-existent (even one might say Pythagorean style) belief either that numbers are prior to actual natural quantities, in some mystical or metaphysical way, or that natural quantities are inseparable from the numbers that we may use to measure or express them.

In pure mathematics we may therefore also confuse the things that *we have invented* called *numbers*, whether the reals or the naturals, with the quantities that they represent. It is true that the *relations* between numbers as we understand them in mathematics are *genuinely objective*, as is the generic form of an infinite iteration process that we might use to create numbers. But they are still part of the process of our mind.

Nevertheless, numbers and natural quantities (quantities in material phenomena) are not the same thing, because what we call a number, and what we encounter as a natural quantity, are distinct objects. They are, however, related, in such a way that we can use numbers to express abstract quantities, in ways that then transfer in application to natural quantities.

What we find is that real numbers cannot perfectly express all continuum quantities that can exist *except as unhalting infinite iteration processes* or IIPs. What we call "irrational numbers" are actually unhalted *number* IIPs that represent quantities on a continuum.

The quantity continuum Q is then considered as the IIP:

$$\circlearrowleft (\infty) \{Q\}$$

and the IIP for producing the natural numbers is

$$\circlearrowleft (\infty) \{N\}$$

and these are distinct objects. But only the latter is itself an infinity of distinct objects. The former is not derived from the infinity of distinct objects, nor is it an expression of it. A natural quantity continuum, or for that matter an abstract continuum, is then not an infinity of distinct objects, but a numberless IIP, even though as a tool we may conceptualise it as with numbers, through the concepts of a "point', and "an infinite number of points".

An alternative to the more usual approach is therefore to look at the *relation* between the various *infinite iteration processes* we have been talking about.

There are infinitely many rational numbers that might seem not to be an infinite process. What we should remember though is that a "number" expressed as a finite number of digits - *a rational number* - when applied to a natural quantity, only ever meaningfully exists as a number in the first instance *as part of* an infinite iteration process, that in this instance has been halted. The process *could* continue to infinity with zeros.

We are painting a picture here of structures of objects that are actually *processes*. At any time we may be considering a structure of objects, as "static" objects, as we might write for example, in the simple structure "$2+2=4$". However, *everywhere* are processes. This particular mathematical truism we just wrote comes from the structure of relations in the first, small, finite part, of the infinity of distinct objects, to which we have given labels called "numbers". The relations in the structure of the infinity of distinct objects is what lies behind and is prior to all relations between natural numbers.

This may in itself seem like a static thing, but behind it is still *process*. And it only arises in our intelligence in the first instance, because of the *process* by which brain function becomes mind.

Furthermore, although we observe brain function as a process happening *in time*, the principle of *process* is not something that necessarily has to happen "in time". An iteration function, or an iteration process, or an iteration function system (IFS) with which we are now familiar as a means of generating naturalistic fractal images, are all processes that in themselves, are *not functions of time*, even though the rendering of the objects they create, through computer processing, takes time.

The Reals and the Continuum

The relation between a real number and the continuum is then the structure

$$[\circlearrowleft \{R\}] \longleftrightarrow [\circlearrowleft (\infty)\{Q\}]$$

where remember the $\circlearrowleft$ symbol by the real number *is* for an infinite iteration, but we may for some numbers consider it halted. The relation essentially says there is a way in which we are applying the IIP for a real number to the IIP for the continuum quantity.

Consider again the primary infinite iteration process

$$\circlearrowleft (\infty)$$

We can envisage this, if we prefer, simply as an *ideal* infinitely turning or cycling wheel, facing towards us, so that the top is rotating towards us, and the bottom away from us. However, this is a crude analogy because we are nevertheless still talking here of an abstract object. There is not any definable point on the wheel for a "beginning" or "end" of a cycle (discussed above as a "mark"). There is no "size" or "frequency" to this iteration on its own.

However, we can still introduce another wheel to the right whose relation to the first, we specify as a *quantity ratio* measured by numbers in what we will design in as 10 cycles to every one cycle of the first wheel. We do not need to add a "mark" object in order to be able to do this. The specified ratio is now a "mark" in itself. We simply specify the *quantity relation* as a number. This is not unlike expressing the *quantity relation* of the circumference to the diameter of a circle object, which is a continuum ratio, as a *number*.

We then can do the same thing again, introducing another wheel, with the same relation to the wheel on its left, so that it will be completing 100 cycles to each cycle of the first. We are clearly producing something like a decimal counter, but without the numbers.

We can also envisage infinitely many wheels. For brevity we can symbolise this theoretical, infinite *device* as, say:

$$\left(\odot \vec{B} \odot \right) \to \infty$$

showing two wheels, whose rotations are related by the number base B, which in this case we are saying is 10 (but could be any base), and the fact that this arrangement extends to infinity.

Now we can see easily that we could also place a decimal point at the far left of this arrangement. The wheels of the device are then set up ready to serve for the displaying of infinite digits after the decimal point, once we take a few more steps in constructing this theoretical "machine". So naturally, we could also start adding wheels to the left of the decimal point, in the appropriate manner, and infinitely, to build up a "machine" for displaying any real number.

As these wheels turn, we can say, they do so smoothly, and so what the "machine" is representing as it "runs" is natural quantities on a continuum, relative to the number base. The very fact that we have a *relation* of cycles measured by numbers, between each wheel and the next (in decimal a 10:1 ratio) means that measurements and numbers are already inherent in the machine. All we would have to do is introduce a fixed "mark" on the circumference of the first wheel, and then all other wheels to the right have an *implied* fixed mark on their circumference, through the fact that their period is related to the period of the first wheel.

Each wheel turns infinitely, and is considered perfectly smooth, and so it constitutes a *bounded infinity* of possible rotational positions relative to the rotational position of the first wheel, all relative to its mark. So in this way a *continuum of quantity* as represented by this machine can be thought of as being encapsulated in an *unbounded infinity* of *bounded infinites*, related to our use of the *natural numbers* to measure cycles or iterations.

It seems at first that the wheels to the left of the decimal (or in principle also some other number base) point, must also extend to an unbounded infinity, like those to the right, if we are to be eventually able to use this machine to express the infinity of the reals (for simplicity we will only be concerned with the positives here). However, alternatively we can have just one wheel to the left of the point if we change the cycle ratio here, and have here a wheel of infinite circumference.

Let us say the number base being used for the cyclical relations of the wheels to the right of the point, is B. This is the ratio of cycles of adjacent wheels, from right to left. Then, essentially, by introducing the big (infinite) wheel to the left, we are relating what is going on in the machine in the wheels to the right, with the *infinite number base*, embodied in the big wheel on the left.

We create finite number-base systems by "counting" the infinity of distinct objects with the chosen number base. All we are doing here is setting up our large wheel to do precisely this, but with an infinite number base.

This "counting" that creates the natural numbers is an infinite iteration process. In our machine, what the wheels to the right of the point are doing, in their rotations, is also an infinite iteration process. It is one that we can now "convert" into any unbounded infinity of numbers to the base B, one for each wheel, simply by dividing up the circumference of each wheel into B equally spaced marks.

Essentially, we are then *relating* the quantities of the continuum represented by the rotational positions of the machine's wheels to the right of the point, to the infinity of the naturals on the wheel to the left of the point. If we were using the decimal number base, then one complete cycle of the first wheel to the right of the point, constitutes a unit of the natural numbers, that by implication are "marked" on the surface of the infinite wheel to the left of the point, in the same way as there is the concept of an "infinite number of points" around the circumference of any circle.

Even at this stage, we can see intuitively that if this operation takes place in time, the large wheel to the left will never move, unless the first wheel to the right is rotating infinitely fast. And if it *is* rotating infinitely fast, the very large wheel, or the infinitely large wheel, to the left will also rotate infinitely fast, unless we can prove that the bounded infinities of "points" of the circumferences of the wheels, are different "sizes".

We might be able to construct, using set theory, an argument that this is the case. But then there will still be the problem of the "ratio" between the cardinalities. Also, this principle applies all the way along the line of wheels. Because the line is an unbounded infinity going off to the right, and because of the rules of relation that we have applied, we will actually never find that the machine's wheels move. However far along the line we go to the right, always, any expected discernible movement will be further off to the right. But if any wheel is moving infinitely fast, then all wheels to its left are moving infinitely fast. This problem of inconsistency does not just apply to the decimal number base as the ratio between the wheels. It applies to any number base.

Is it that we made a false step in specifying the ratio between the wheels, without a "mark"? It is not, because the same problem arises even if we begin with all marked wheels. The problem arises from the relation of the unbounded infinity of wheels to the right, with the bounded infinity of not just the large wheel, but *any* of the wheels. The problem exists in the apparent relation between what we have *conceived* as an unbounded infinity, and a bounded infinity. This is something addressed in the book *Structures and Morphisms*.

Essentially, what we have just done is to partly construct a theoretical machine for measuring a quantity on a *quantity continuum*. It represents any continuum quantity by the positions of rotation of its wheels. After applying the marks to all the wheels for the number base we are using, then to complete the construction neatly, we would need to arrange for each wheel to only spring into its next marked position each time the wheel to its right has completed a rotation. The machine then has a "readout" that corresponds to actual real numbers. Real numbers are strings of definite digits, and never strings of positions somewhere in between two digits. To complete our neat building, we would also need to replace the infinite wheel to the left to the point, with an unbounded infinity of wheels.

What we have now done, is to get rid of bounded infinities. Each wheel now represents a finite number - according to the number base - and the only infinities in our machine are now the unbounded infinities of adjacent wheels left and right of the point, and the available infinity of iterations (cycles or rotations) of each wheel. We can now relate the machine to any continuum quantity, and it will present the measurement of that quantity *as a real number*.

Now, the only problem is the question of how far along the line of wheels to the right we want to look, when the continuum quantity being measured corresponds to an irrational number, because in practice we can never look infinitely far along. This corresponds to the way in which we must terminate the digits after the decimal point, for an irrational number.

What we are doing, is applying this number quantity machine, to a continuum quantity. We do this *whenever we express a continuum quantity as a number*. A continuum quantity and a number are *two distinct objects*. However, if we now remove the marks, and allow the wheels once again to rotate smoothly without the spring effect, then the machine is still capable of being applied to a continuum quantity, and furthermore, its state can exactly match the continuum quantity *even if there is only a finite number of wheels after the point*. In fact, we can get rid of all the wheels, and have just one wheel. Continuum quantities exist without numbers, and are always related to the infinite continuum.

However, without marks, there is no measurement, and no number. The position of the wheel on the continuum it is measuring is itself a continuum quantity. It is no longer a number object, but a continuum quantity object.

We now have the relation between the quantity M defined by the machine, and the continuum quantity Q, as:

$$\circlearrowleft (\infty) \{M\} \leftarrow \stackrel{=}{-} \rightarrow \circlearrowleft (\infty) \{Q\}$$

We can envisage that if we go back to our marked machine, with wheels that spring into position, that we could apply it to an infinite continuum. But if we did, all the wheels would be infinitely turning. They would not all be at "9", or "0", or any other number. Rather, the machine would not be halted. It would be a *process*. Specifically, it would be an IIP.

Summary of the Object Relation Between the Reals and the Continuum

If we understand not only the irrationals, but *all* real numbers, to be either infinite iteration processes (IIPs), or a halted part thereof, as we have seen, and we also understand a continuum quantity as the IIP of a *continuum quantity* object, then there is a *clear relation* between the real numbers and any continuum quantity. With thus jettison the notion of a continuum quantity as "numbers", and see the relation between number quantities and the continuum quantity.

That relation is then:

- They are then both viewed as IIP constructs, or if we like, in terms of Object Theory, *structures* of IIPs.

- A *continuum quantity* and a *number*, are two distinct objects, and remain two distinct objects, even though the number object can be a perfect measure of the continuum quantity object.

We cannot show that two distinct objects - such as the "set of all real numbers", and a *continuum*, are *not* distinct, unless we are demonstrating a fallacy or a paradox. What is genuinely objective is neither a fallacy nor a paradox. Fallacies and paradoxes arise only as limitations in our understanding.

11: From Zero to Infinity

W hat is the mathematical "division" object? The idea of "divided by" or what many mathematicians call some number "over" some other number, is the "ratio" object. A "ratio" between one quantity and another, is a specific kind of *relation* between the two quantities. It is a *relation* that is itself a quantity object. We evaluate it using numbers. So first, let's look at numbers objects, and the process of multiplication.

Numbers stand for quantities *of* something, and in pure mathematics they can represent quantities of abstract units. The *natural* numbers, specifically, as distinct from real numbers in general, refer to quantities of distinct objects, even if these objects are entirely abstract and only implied. In accordance with Object Theory, objects in any case don't have to *be* anything identifiable in any way other than that they are an object. They can be just abstract, distinct objects. The criterion here, is that they are *distinct*.

So "6" refers to 6 *instances of distinct objects* in the Object Theory context. These objects can be identical, but they must be distinct. In arithmetic, for example, they may be "units", and all "units" are identical.

So a "natural number" such as "6" implies a *structure* (a network of related distinct object) where the *quantity* of objects is what we in our intelligence call the *number* of objects. We could call this structure of distinct objects, the "internal" structure of the number object, such as "6".

What happens in the case of zero? What structure does it imply? Zero is still an object. However, as a mathematical object, an object in the context of mathematics, *until we relate it* to other objects, it is an isolated object with no inherent relation to any other number object. It also has, in itself, no "internal structure" other than a structure of brain function through which it is conceived.

Multiplication

A multiplication operation is an object. In Object Theory it is a structure that is a *process*. A process that may or may not be infinite, or non-halting. In the structure "6×2" we are talking about the finite or halting process in which the "inputs" are objects "6" and "2", and the "output" is the object "12".

We don't have to be using decimal as the symbols for the structures. The structures stand in their own right. However the counting process that we use, determined by what we call the "number base", is a process in its own right, which we then relate to these structures.

There is, as it were, a "base level language" consisting of structures of distinct objects, which we then conceive as "numbers", which in turn then get "translated" into the "higher level language" of the written number symbols and symbol strings.

A mathematical operation such as multiplication, is then a process object that is applied to the "higher level language" in order to represent structures in the "base level language" of structures of distinct objects. The analogy to computer programming is not merely accidental.

The structure "6×0" stands for the application of the *counting process* to 6 instances of the zero object, and the structure "0×6" applies the *counting process* to zero instances of the "6" object. Either way, the rules of the *counting process* results in the number object "0", which stands for "no quantity" because it has no"internal" structure of distinct objects.

Irrational Quantities

Here, of course, we have been dealing with natural numbers. What happens in the case of real numbers in general? Now, the ordinary expectation is that we are dealing with a continuum. As we have said, the continuum, as an abstract object, is a process object. It is the structure to which we then choose to apply the process of real number generation. We looked at this above.

We can of course write any single symbol, to "stand for" any irrational number, like just S for example. But if we want to talk about the irrational number expressed using the symbols of a counting process, then it can only be expressed with an infinite process. And of course, we cannot literally write that process as an object that is not a process, such as a finite number string. We can only symbolise it, with a simple such as π.

We can consider the irrational quantity structure as a partition. This is a specific kind of *relation* between a quantity N and a quantity D, the kind of relation that we called a "ratio" as N/D. These quantities we are talking about are not *numbers*, because we are not yet talking about number objects. They are *quantity* objects. And we can put such an object together with the continuum object to make a structure that is a quantity of continuum. The obvious everyday example of this would be the length of the circumference of a circle, and the length of the diameter of the circle. Each is a quantity of continuum. Notwithstanding that we may be conditioned to think about this as an "infinite number" of discrete "points".

Now we may quantify these length objects with numbers, but they exist as objects in their own right. They are part of the phenomena of mind and experience and comprehension that is constructed through our brain function. What we are doing here is representing an aspect of the principle by which that brain function becomes our experience of the world and our comprehension of mathematical structures, in terms of structures of objects.

What this amounts to, at the same time, is a *thought structure*. We are in essence circumventing the need to understand how brain function construct our experience of the world, and our comprehension of mathematical structures, by going to the root of the way in which we think objectively.

In the case of the circle on a Euclidean plane, any self-consistent measuring system using numbers, that we apply to these two quantity objects, the lengths of the circumference and diameter, will come up with the same ratio between the two, because that ratio is inherent as a specific kind of *relation* between the two length objects.

However, our long history of neuroplastic configuration in the way we think about this (passed down generations through teaching and learning) drives us to still effectively revere some of the ideals of the Pythagoreans, in that we believe the length objects to derive from numbers, as if numbers are some kind of primal, transcendental reality, independent of the intelligence we are being.

We can now normalise what we are looking at here, by designating one object, say the diameter of the circle, as the "unit". Clearly, as you will discern, where we are going with this is towards the irrational number π.

The structure of this "irrational" quantity that is the ratio of the circumference to the diameter, we can only express as a number, by using a number object that is an infinite iteration process. In contrast, the original undefined abstract *relation* between the circumference object and the diameter object (or if you prefer between their respective attendant length objects) is not usually defined as a process. This gives rise to the difference between what geometry can express, and what arithmetic can express, that the ancient Greeks were well aware of. Whilst Pythagoras held numbers to be fundamental, or so we are told, the ancient Greeks in general held geometry to be superior to arithmetic.

Counting

A counting process is a process that either terminates or it does not. The infinite iteration process or IIP of the naturals, does not terminate. All finite numbers are terminations or haltings of the infinite process. They are, in essence, a morphism of an *infinite process* object (which is the structure of the counting process), into an object that is not a process.

Similarly, where we are looking at numbers that are said to "go to infinity", what we are looking at is a morphism from something that seemed to us not to be a process, into an infinite process. The axis of a graph is a good example. Anywhere *on* the axis is a finite number. However, there is no point *on* the axis, that is "infinity". There are only numbers *on* the axis, and infinity is not a number. The axis is *defined* by numbers. The infinity that we say it "goes to" is not the axis, but the process of which the axis is a part.

What happens when the axis is used as the basis for describing an infinity, such as where we say a graph plot "intersects the axis at infinity", is actually that there is an *apparent* morphism from what we were treating as a non-process, into the infinite process.

The morphism itself is an artefact of the way we have been treating the situation. For example, if we plot $y = 1/x$, then where $x = 0$, there appears to be a morphism of the axis and the plot, to the infinite process of the real numbers. This morphism is not something that just suddenly appears at $x = 0$, in a thing that we call a "singularity". Rather, it is that at $x = 0$ we no longer have the choice of treating what we are looking at *as though* it is not a process. In fact, the axes of the graph or already part of or derived from infinite processes, and the plot of the graph is an infinite process. It's just that we usually look at only portions of these things, in which we can treat them as though they are not processes.

When we do this, we encounter morphisms. Whenever we encounter morphisms, which are usually called such things as "singularities", it is an artefact, and a symptom of the way in which we have been comprehending. It's not that we shouldn't comprehend in the way that we do, when that is convenient, rather, it is that we should understand what we are doing. And when we want to go beyond the limits of the way in which we are doing things, we need to understand mathematical structures in terms of infinite processes.

More generally, if we are talking about all numbers, or the real numbers, rather than just the natural numbers, then if we are just talking about a single number, extracted from the infinite process of generating the numbers, we may already be talking about infinite iteration processes, because even a single irrational number, is an infinite iteration process.

Given that all valid mathematical structures must be able to "translate" into structures of relations between numbers, then what we are looking at more generally, in all mathematical description, is the relation between objects (structures) that are infinite iteration processes (the processes of generating numbers), and objects (structures) that are not.

In short, the most interesting part of all mathematics is the part where it breaks down. The part that we ordinarily call "singularities". We have developed set theory, and a set theory "understanding" of infinity, in which we believe the mathematics we are using has no longer broken down. But that is a belief, rather than something that is genuinely objective.

The Division Operation

We just looked at the irrational quantity as the object structure of a partition. The division operation is interesting. We can always view it in terms of partitioning. Let's say we are talking about a division operation N/D. The operation has to be the same no matter what numbers N and D are. Therefore, N and D do not stand for specific numbers. Rather, the division operation, or ratio, is a specific kind of *relation* between two objects, N and D, which represent the infinite iteration process of the real numbers.

The essential task is then to find, *at any halting* of both N and D, what number R we need to multiply D by, in order to result in the N. Again, we apply a *process* for doing this. And again, depending on N and D the process may or may not halt. If it does not halt (as would be the case when R is an irrational number), then it is an infinite process.

Singularity

What happens when D is zero? As is well established, it is not the case that we find a *number* called "infinity" that we can multiply D by, in order to arrive at N. Because D is already zero, there is no number R such that DR = N.

However, it *is* the case that the ratio R "goes to infinity", as we say. The term DR is no longer properly defined, as its literal meaning would now be "0 multiplied by infinity", as if infinity is a number. Essentially, though, what the equation "says" still holds in a pedantic literal sense, because DR is no longer defined, which means that it can in effect equal *any* number, including N. Correspondingly, N can be any number, without affecting R, as long as D is zero.

It is the fact that in the division process, D is in the denominator position, that results in the numerator N, no matter what it was, "escaping to infinity" when D is zero. We said that the division process was essentially the process of finding what number R we need to multiply D by, in order to result in the N.

The simplest structural way of creating a division process, is, *for any value of* N, to use an iterative algorithm for homing in on likely real number answers with the condition built into the process that the process halts when the number generated fulfils the required condition: *halt when the number generated multiplied by D results in N.*

When N=0 the result is that the applied process does not halt. It *appears* that R undergoes a morphism from being a number, to being the infinite iteration process, when D is zero. But at what point does this morphism happen?

This is equivalent to asking at what point do the real numbers of diminishing size as the value of D, actually become zero? Of course there is no such point, because we are not looking at points to begin with, we are looking *processes*. The idea that morphism just happens when D is zero, is mistaken. The entire structure is always undergoing a morphism, because the entire structure is a *process*.

If we write, algebraically, R = N/D, then unless we are using a convention where we are talking about just one specific number, for each of those "variables", it is the case that this is algebraic symbolism for a structure that is a *process*.

The graph R = 1/D

We can illustrate this same thing in another way in the form of the two-dimensional graph $y = 1/x$. Here, we have simply fixed the numerator at 1 for convenience. We usually say that if $x = 0$ then y "goes to infinity". What is actually happening is that at $x = 0$, y is no longer a number, but rather, has become a different object - an *unhalted* IIP for finding the value of y.

We usually say that the graph plot *intersects* the y axis. That's because we imagine there is some "point" on the axis that is "infinity". As if infinity is a number, or a point on the axis, that itself consists only of numbers. But in fact, there is no *point* on the axis that is "infinity". The axis itself is a *continuum quantity* IIP, to which we then apply numbers. What happens is that the plot of the graph at $x = 0$ has *become* the same object that the y axis becomes "at infinity", which is simply an unhalted IIP.

What happens to x when y is zero? Again, it is tempting to say that x is "infinity". But once again, infinity is no more a position on the x axis than it is on the y axis. When y is zero, x is the unhalted IIP of the x continuum.

These are not really *singularities*, in the sense that they are an *exception* to what is defined, or going on. Rather, if we find them to be *singularities*, or exceptions to what is defined, or understood as going on, then that is because we are not fully understanding what is happening to begin with. We are trying to understand things that are structural *processes*, as if they are something other than processes.

0/0

What happens when both D and R are zero? This is conventionally referred to as "undefined", because it *is* undefined, in the conventional way of understanding what is going on. We can see it if we look at the three-dimensional graph of $y = N/D$ over the domain of real numbers both negative and positive for N and D. When both N and D are zero, the surface of the graph coincides with the y axis.

Of course we can never actually *see* "what happens at infinity" in the graphs, because the y axis never actually goes to infinity in practice. However, it is relatively easy to get the idea of "what is happening" in the three-dimensional graph, compared to the two-dimensional graph.

Again it is not that the surface of the graph *intersects* the axis as might be commonly said. Rather, what happens is that when N and D are both zero, there is *no longer a distinction* between the two objects, the surface, and the axis. There is only one object, which is the IIP of the y axis continuum.

Also, again, it is not that there is some sudden, unaccounted for change, between something that is not a process, into something that is a process - the infinite continuum that the axis represents - when D hits zero. The x axis itself - any number on it - is a number applied to a continuum IIP.

The graph plot or surface is also also *already a process*, that again, we can only partially represent as if it is something other than a process, by drawing a *part* of it. What is happening at $D = 0$ is that one process has become another. The process that is the surface of the graph, becomes the process that is the y axis continuum. This *morphism* is smooth and continuous. There is nothing strange or out of the ordinary about it. It is still part of the structure of objects just as is the surface of the graph elsewhere.

Essentially, in terms of objects and structures, the "singularity" at $D = 0$ is no more an exception or singularity, than say, the minima or maxima of waves. It is simply a kind of "turning point" in the continuum structure, and all continuum quantities are always structurally related to infinity, through the continuum IIP.

Discussion

In any process, we can take some point in it, or halt it at some point, even if it is an infinite process, and then find something that is apparently not a process. If we were to generate the Mandelbrot set image, we could halt the process at any point, and there would be just such an object, that is not a process. But if we wanted to examine further, if we want to "zoom in" further, then we have to start up the process again. And yet, at the same time, the Mandelbrot set is not something that "takes time" to exist, just because our computer may still "take time" in order to produce an image of it. It is already complete and finished, as an infinite object. But it remains a process object.

The graph plot $y = 1/x$ is a process, and the plot of the Mandelbrot set is a process. With either, we can halt the process anywhere, to see a "partial picture" of the whole object. When we in practice plot a graph, we only ever plot part of it. The same is true even for the Mandelbrot set image, whose geometric infinities all converge.

When the process of a plot or graph surface loses its distinction from one of the axes, then what we have seen is a *morphism* from one structure to another, different structure. In the graphs we have been talking about, the graph plot or surface undergoes a morphism as x approaches zero. When x is zero, the morphism is complete. At what point does the morphism begin? It begins wherever we start plotting the graph or surface. The very nature of the graph or surface is that it is a process object that undergoes a morphism into another object.

There is nothing strange or even "undefined" about what is going on. What is going on, is processes. One of our greatest impediments to our being able to correctly conceive infinity in the context of mathematics, is in trying to conceive it as something other than a process.

This is not unrelated to the idea that what we are thinking about, in mathematics, consists of something other than processes of thought. It does of course, actually consist of processes of thought. Out of these processes of thought, the endeavour is to reason objectively, with mathematical objects. However, even when what we are representing mathematically is genuinely objective, our understanding of it, and our reasoning in that way, is still a construct of brain function, which is a process in time. A process that in its relation to our mind and our world, and the way in which it creates our experience of both, we have not yet come to understand in science.

So firstly, we must fully acknowledge the nature of *process* as a structure, and indeed, one that can be infinite, and not at all dependent on time. Except, that is, through the *time* that is a part of the process of thought.

Secondly, we need to acknowledge the ubiquitousness of processes everywhere in both nature and mathematical reasoning, and even that objects we invariably consider in other ways, are processes or parts thereof.

Thirdly, we must recognise that all our object oriented understanding is process based - because it is based in the processes of thought, which are themselves, part of the processes of time.

12: IIPs and Infinity

To make the statement that "There are an infinite number of points around the circumference of a circle", is to present a structure. It is a structure of objects, of concepts, of comprehension, a thought structure, a concept structure.

In Euclidean geometry we often talk about points and lines. A "point" is an object, and a "line" is an object. Another thing that is often spoken, is the phrase "an infinite number of...". Let's reduce that to the object "an infinite number".

Is there really such a thing? From what we have previously said, there is not, not if the object "number" is considered in the way that it is conventionally considered. That is, as something other than an infinite iteration process, or a part thereof. There is no number object that is infinite. Just as "infinity" is not a "place" or "position" on an axis, a thing that relates position to number objects.

If we use commonplace terminology, then in Euclidean geometry we have such ideas as "A single point is a line of zero length". And also "There are an infinite number of points around the circumference of a circle". We also have "there are an infinite number of points in a line".

These are structures, literally concept structures, elaborated from from a more fundamental and simple structure. If we carry on like this, then we have the following:

1. No points implies no line (which seems fair enough).

2. A line of zero length has one point.

3. A line of *any* finite length has an infinite number of points.

4. Any *finite* number of consecutive points is a line of zero length. This also subtly implies that there is such a thing as a *number* that is infinite.

5. An infinite number of consecutive points is a line of any undefined length, greater than zero.

In fact, of course, "points" do not belong to lines in Euclidean geometry, but rather, lines belong to points. Points are nothing more than positions in the system of coordinates. A straight line is essentially, when we strip away elaborating concept-structures, an "interval" in the system of coordinates. So let's restate this in another way:

Table Row No.	Number of Points	Relational Direction	Interval
1	0	>>	0
2	1	<<>>	0
3	Finite n	>>	0
4	Infinite	>>	Undefined
5	Infinite	<<	$L > 0$

This rather strange-looking set of relations arises naturally from the *structural relation* between the objects *zero*, *one*, and *infinity*.

Row 1

First, the object "0". We are saying here that this is the "number of points". If it is meant to be part of a number system that can be applied to specify a position in a quantity continuum, which in Euclidean geometry it indeed is, then as a *structure*, we can say it is a halting point in the infinite iteration process:

$$\circlearrowleft (\infty) \{Q\} \quad (1)$$

where Q is the quantity continuum. However, we are also saying that it is an object in the counting system for the infinity of distinct objects, which in this case, we are instantiating as distinct things called "points" on the Euclidean plane. In other words, for example, what we are calling "0" here, is also a digit in the decimal counting system that we would use to count distinct points.

The decimal counting system that we can use to count the number of points in the interval is the same as the infinite iteration process for producing the natural numbers (on the infinity of distinct points):

$$\circlearrowleft (\infty) \{N\} \quad (2)$$

So then, "0" here is the object that stands for the initial condition or "starting condition" of both (1) and (2).

The interval in the co-ordinate system is the corresponding quantity Q on the co-ordinate system. This is the halting point in

$$\circlearrowleft (\infty) \{Q\}$$

and this halting point we have already defined as "0". So the interval is "0", and the "number of points" is "0".

Row 2

Here, the "number of points" is the number object "1". The same conditions apply as for Row 1, but we are now saying the number of points is "1". That means the infinite iteration process used for counting points:

$$\circlearrowleft (\infty) \{N\}$$

halts at this number. Meanwhile, on the continuum of quantities of the co-ordinate system, nothing changes, because "1 point" is by our own definition a single coordinate position, which is a position on or a state of

$$\circlearrowleft (\infty) \{Q\},$$

and hence there is no interval, and the interval is "0".

Row 3

Now here it is the case that

$$\circlearrowleft (\infty) \{N\}$$

halts at any finite number. The reason this has no effect on

$$\circlearrowleft (\infty) \{Q\}$$

is that we are only here applying

$$\circlearrowleft (\infty) \{N\}$$

to these objects called "points" that we also at the same time happen to be *envisaging* as existing in the

$$\circlearrowleft (\infty) \{Q\}$$

infinite iteration process of the co-ordinate system. However, we have not shown any actual structure of *relation* between

$$\circlearrowleft (\infty) \{N\} \text{ and } \circlearrowleft (\infty) \{Q\}.$$

So no interval on

$$\circlearrowleft (\infty) \{Q\}$$

has been created, it is untouched by the state of

$$\circlearrowleft (\infty) \{N\}$$

due to the fact that it has no structural relation to it.

Row 4

Now we are saying there are an "infinite number" of points. What this actually means is that

$$\circlearrowleft (\infty) \{N\}$$

which is an IIP, does not halt. Since it is not halted, there is no halting position of

$$↻(\infty)\{Q\}$$

that even *could* be related to it, to make a structure that is not a process. Hence the interval on

$$↻(\infty)\{Q\}$$

is both undefined, and may also be infinite.

Row 5

Here we are looking at the reverse relation. We *begin* by saying that there is a finite interval greater than zero. This means that

$$↻(\infty)\{Q\}$$

has halted. We must now apply

$$↻(\infty)\{N\}$$

which is for counting distinct objects in the infinity of distinct objects, where in this case we have chosen those objects to be "adjacent points" that we envisage are on

$$↻(\infty)\{Q\}.$$

What happens? Simply,

$$↻(\infty)\{N\}$$

does not halt, because no such points exist on

$$↻(\infty)\{Q\}$$

due to the fact that it is not an object that includes these supposed "points" as part of its structure.

Row 2 again

There is one last relation that we haven't mentioned yet. This is the reverse relation in table row two. This time we *begin* with an interval of "0", which means we instantiate

$$\circlearrowleft (\infty) \{Q\}$$

in the halted condition, and take this to be a starting condition. It exists as a position on the infinite continuum that

$$\circlearrowleft (\infty) \{Q\}$$

would otherwise create. Since every position on the continuum is designated, according to our presumption, to be "point", then when we apply

$$\circlearrowleft (\infty) \{N\}$$

to the halted

$$\circlearrowleft (\infty) \{Q\}$$

under the presumption that there must be a point there, we naturally find its process to halt at "1".

What we have done here, in all these examples, is separate out the components of our thinking and reasoning through the concept of "points", into their structures.

Speculating about "points" on "lines" or as things in co-ordinate systems, in general, is precisely the same as assuming that quantities on a quantity continuum can be treated as the same thing as the real numbers we may use to describe them. This is a mental presumption rather than a mathematical fact.

Such things are schemes of understanding, ways of learning and conceptualising, that usually are useful and can be successfully used as intellectual tools. However, they are not an intrinsic "truth".

It is also perfectly objective to say, if we want to, that there is *not* an infinite number of points around the circumference of a circle, if you choose not to use that scheme of understanding. The points are not *genuinely objective*, they are merely collectively agreed objective tools used as a means to an end, in a consensus network.

A 3D Cartesian system of co-ordinates can also be viewed as a structure of 3 infinite iteration processes, one for each axis, as:

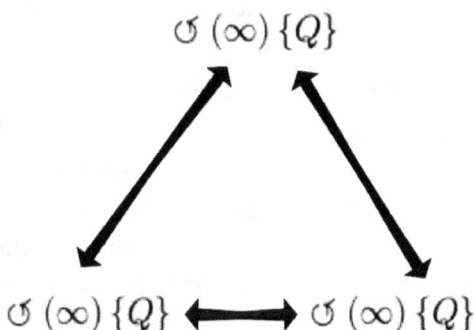

$$\circlearrowleft (\infty) \{Q\}$$

$$\circlearrowleft (\infty) \{Q\} \longleftrightarrow \circlearrowleft (\infty) \{Q\}$$

This structure of three objects is constructed from three distinct instantiations of the object:

$$\circlearrowleft (\infty) \{Q\}.$$

In the representation above, the structural *relations* shown by the arrows mean that the objects are distinct, even though they are identical (the fifth premise). In this particular case we usually call this distinction *orthogonality*.

What if we change the relation? Let us for brevity rename these three objects (each of which is an infinite iteration process), as z, y, and x.

Let us also define a process object "/" that is a ratio relation, with two inputs and one output. The output is the ratio relation of the two inputs. However, let us say that it does not express this relation in its output, using a numerical process that utilises a numbering system, such as

Page 101

$$\circlearrowleft (\infty) \{N\}.$$

Rather, its two inputs are simply quantities, that are from a quantity continuum. For example, if we were to construct a circle of a diameter D, the circumference length would be what it turned out to be what it turned out to be, through the operation of rules and principles, or laws, that we encapsulate as a mathematical formula, which is a structure of number relations.

However, although we might in capsulate these principles in this way, the process of creating the circle, the manner in which the circumference forms itself, does not involve the doing of any mathematical calculations. It just happens in the way that it does.

Similarly, if a planet happens to orbit a star in a perfect circle, let's say, it does not do so, as a consequence of working something out, as a matter of relations between numbers. It does so as a consequence of the way structures of relations between natural phenomena work, which we then come to understand, through the invention and use of numbers.

We may like to think that the numbers are inherent in the phenomena we see, and so if we see seven standing stones in a field we think this is a demonstration of a number, specifically, the number 7. What it is, is a specific structure of plurality, one of the many that occur in nature, that we then *label* "7", or "111", and so on, according to the number base we choose.

In this way we come to represent principles and laws inherent in plurality in nature, as numbers and number systems, and then see nature as embodying numbers.

However, all this happens in the intelligence we are being. The way in which we see numbers and number structures in nature, is actually the way in which our own intelligence is coming to understand natural structures of relations between distinct objects in the phenomena of our world.

It is the easiest thing in the world, once we learn to count, and to play with numbers, and their relations, to see this aspect of our mind, in the world, and then to presume that numbers are something transcendental and separate from our mind, behind the appearance of the world, because we don't see the symbols and the formula themselves, floating around in front of us, as part of the natural phenomena. *And* because we assume natural phenomena to be something other than the start of our own mind.

This idea is the then further reinforced when we find that it exists in ideas associated with the ancients. There is nothing quite like the association of an idea with the ancients, and also with the idea of "lost knowledge" which dominated in the European intellect during the Renaissance, to give it a feeling of legitimacy. But really, mathematicians, of all people, should be above this sort of thing. That is not to deny that there are things that are transcendental. But we need to have more knowledge of the nature of our own intelligence, and the intellect we are using, before we can come to know what they are.

All over nature we find continuum quantities interacting with each other, without any explicit demonstration of "numbers" or "number like" plurality of the kind we can also find in nature, just in the plurality of distinct objects, such as standing stones in a field.

We can also represent relations between continuum quantities, if we want to, without recourse to numbers. This is why the Greeks held geometry in the highest esteem. But today, we consider numbers to be pre-eminent, and perhaps somewhere in our psyche, we even still consider them to be somehow fundamental in a transcendental and metaphysical way. The error there, is not in the perception and understanding of numbers, which is essential to our understanding, but in the presumption that anything we think or understand, including numbers and mathematical structures, is about something *separate from* the mind we are being, and *separate from* the brain function through which our mind is arising.

So we can define the "internal" structure of this process object "/" as:

$$[\mho_1\,(\infty)\,\{Q\}] \longleftarrow \overset{\text{ratio}}{} \rightarrow [\mho_2\,(\infty)\,\{Q\}]$$

where the two sides here, left and right, are the two inputs. If we say the output is the quantity continuum object

$$[\mho_3 \, (\infty) \, \{Q\}]$$

then we could represent the whole structure as:

$$[\mho_3 \, (\infty) \, \{Q\}] \equiv [\mho_1 \, (\infty) \, \{Q\}] \longleftarrow \overset{\text{ratio}}{\longrightarrow} [\mho_2 \, (\infty) \, \{Q\}]$$

All we have done is take the *structure* of the 3D Cartesian system of coordinates that we looked at above:

$$\mho \, (\infty) \, \{Q\}$$

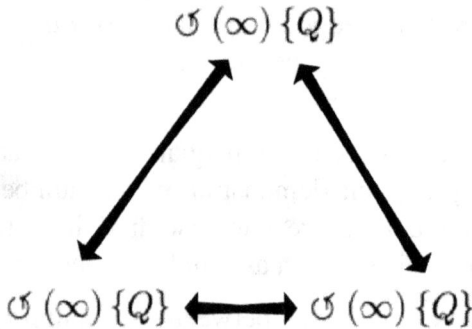

$$\mho \, (\infty) \, \{Q\} \longleftrightarrow \mho \, (\infty) \, \{Q\}$$

and changed the *relations* (which we specify on the *relation* arrows) between the three infinite iteration objects.

This is more usually expressed simply by, for example, having:

$$[\mho_3 \, (\infty) \, \{Q\}] \equiv z$$

$$[\mho_1 \, (\infty) \, \{Q\}] \equiv y$$

$$[\mho_2 \, (\infty) \, \{Q\}] \equiv x$$

and saying

$$z = y/x$$

Page 104

and then considering z expressed by the surface of the *graph* in the *space* of the system of coordinates.

However, the presumption here is always that x, y, and z are *sets* of static and distinct *number* objects. It is not even usual to think of them as number objects, but simply as *numbers*.

So following what we have been saying, the set of relations

$$[\eth_3 \, (\infty) \, \{Q\}] \equiv z$$

$$[\eth_1 \, (\infty) \, \{Q\}] \equiv y$$

$$[\eth_2 \, (\infty) \, \{Q\}] \equiv x$$

are actually chimerical. That is to say, if x, y, and z are each *sets* of static, distinct number objects, that are other than processes, then each one of these relations is falsely stating that:

$$[\eth \, (\infty) \, \{Q\}] \equiv [\eth \, (\infty) \, \{R\}]$$

in which the object:

$$\eth \, (\infty) \, \{Q\}$$

is the quantity continuum in question, and the object:

$$[\eth \, (\infty) \, \{R\}]$$

is the real numbers. And this brings us back to the fact that the real numbers R, and the quantity continuum Q, are two distinct objects, that have different structures.

In the previous chapter we talked about this, and particularly about the "singularity" at $x = 0$. We illustrated how this "singularity", when we view the components as IIPs, is not like an exception in the way the relations between the quantities otherwise work. On the contrary it is just another ordinary part of the process.

In the conventional conceptualisation of $z = y/x$ the function is commonly conceived as an object that exists in the space whose dimensions are x, y, and z.

As a structure this conceptualisation is of two distinct objects, the function object $[z = y/x]$ and the space object $[x, y, z]$. We might represent the structure as:

$$[z = y/x] \longleftrightarrow [x, y, z]$$

The function space consists of the structure:

$$\circlearrowleft (\infty) \{Q\}$$

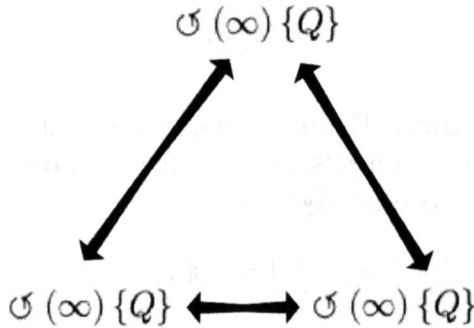

$$\circlearrowleft (\infty) \{Q\} \longleftrightarrow \circlearrowleft (\infty) \{Q\}$$

where the three axes are distinct, and we say that their relation to each other (the arrows) is that they are orthogonal.

In the structure of $[z = y/x]$ considered as:

$$[\circlearrowleft_3 (\infty) \{Q\}] \equiv [\circlearrowleft_1 (\infty) \{Q\}] \xleftarrow{\text{ratio}} \rightarrow [\circlearrowleft_2 (\infty) \{Q\}]$$

the three IIPs now have the structure:

$$\mathcal{O}_3(\infty)\,\{Q\}$$

Identical

$$\mathcal{O}_1(\infty)\,\{Q\} \longleftrightarrow \mathcal{O}_2(\infty)\,\{Q\}$$

Ratio

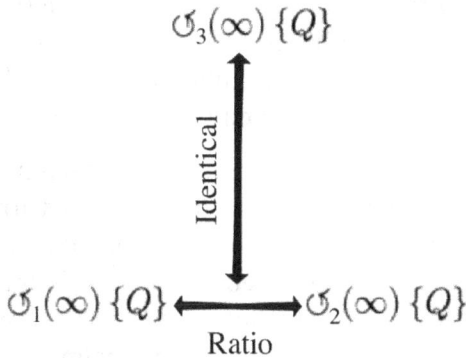

So if we consider functions and numbers as structures of IIPs then the concept of the *space* becomes redundant. Rather, what we are looking at is just structures of relations between IIP objects. In the natural relations between quantities as they appear in finite nature, we can then account for the existence of implied infinities as arising fundamentally from the principle of the infinite iteration process.

Our ability to handle numbers, to count, and to measure, and to determine algebraic structures that ultimately must stand for number structures, all rests on the principle of the IIP.

If we understand the IIP as an object in its own right, a principle in its own right, before we furnish this object with any particular properties, then we are thinking along the right lines.

We then find that we can apply the principle, apply this object, to another fundamental object that exists in our comprehension, which is the infinity of distinct objects. We are not claiming that this structure we are calling the infinity of distinct objects is *genuinely objective*, according to the definition of *genuinely objective* in Object Theory. But it most certainly exists as a n objective concept, an *object*, of thought, that we are employing when we do all our intellectual reasoning about numbers.

It is only by applying the IIP principle to the infinity distinct objects that we create what we call the natural numbers, or a natural number counting system.

We then also apply the IIP principle to the creation of what we call the real numbers. And there, sometimes the IIP halts, and becomes a finite iteration process, as in the case of rational numbers. In the case of the irrationals, the IIP does not halt, and remains an IIP.

In the practical case of making actual measurements of quantity continua in science, even if we are measuring an irrational quantity, the IIP that we are using in order to make the measurement, has invariably already been instantiated *in practice* in the form of a finite iteration process for the production of the measurement figures. We can only ever express to a finite number of decimal places.

So in classical science we are already bound to iteration processes, because we are using numbers in order to quantify both continuous and discontinuous phenomena. It is not therefore such a great leap to see that *natural quantity continua themselves*, can be considered in terms of the numberless IIP object.

Space, Time, Objects, and Brain Function

Relations between measurements of phenomena in science are invariably considered as existing in "spaces". Even in the case of discontinuous phenomena such as the quantum state, we are invariably talking about a space of some kind, for example, infinite dimensional Hilbert space. "Space" itself, however you want to view it, and of whatever kind we are talking about, is itself an object. However, as is the case with any object in Object Theory, that in itself does not necessarily mean that it is genuinely objective.

On this same basis, even what we call "space" and experience as "space" in the everyday world, is an object, in Object Theory. The same is true of time. And, indeed, as it happens, in their combination and extension to the universe in general, spacetime is an object, in the theory of relativity.

Empirical verification of the theory of relativity shows the spacetime object to be genuinely objective. However, none of this means that the mathematical tools we use for describing the relativistic behavior of natural phenomena are something with status other than mathematical tools. In other words, the central object of relativity theory is the mathematical manifold Einstein called *spacetime*, and this is widely talked about and described as though it is genuinely objective as an object other than a mathematical tool for describing the relativistic behavior of natural phenomena. This comes from the fact that it replaces Euclidean space and time, the everyday space and time that we actually experience as a sentient beings.

We should not forget that we are sentient beings who only have our experience of being, and our experience of space and time, and only have this intellectual understanding in the first instance, as a construct of brain function.

The fact that something is *genuinely objective*, meaning (according to our definition) that it is not dependent on any particular mind or network of minds (or brain functions), does not automatically also mean that it exists as an object that is not dependent on brain function in general, for it to be knowable. It does necessarily not mean that it is something to which brain function and brain function's means of giving rise to mind and reasoning, is irrelevant.

We could ask, does the object we conceive, for example, as infinite-dimensional Hilbert space, actually occur in nature independently of the mathematical constructs that exist in our understanding (which is a construct of brain function) for which it was contrived in the first instance?

The metric of relativity that a whole network of scientific minds now refers to *spacetime*, is indeed a metric, whose measure is the *spacetime interval*, one that is important because it is invariant with respect to other systems of coordinates. But it is also perhaps *conceived* as something more. Because after all, it is an "extension" of the concepts of space and time which we also use metrics for, in Newtonian physics, space and time that we actually *experience* as part of the nature of our world *as sentient beings*, whose experience of mind, world, and intelligence, arises through brain function.

We cannot continue to simply ignore the question of the relation between the way brain function creates our experience of being, mind, and world, and the ways in which we are describing the physics of our world.

The way we experience our own existence is in terms of space and time. Because we first consider the world in which we experience our existence, to be separate from us, we consider space and time as independent of the intelligence through which it is known. However, this is psychological position, not a scientific fact.

Our psychological position that exists behind our intellectual reasoning activities has far reaching influence. Consider the case of a pure mathematical function such as the one we talked about above, $z = y/x$. There, describing the function in terms IIP quantities, and rejecting the idea of *numbers* having some "transcendental" style existence, it's easy to see the space in which we ordinarily envisage the function to exists, as something that is just a conceptual extension from the nature of the function itself, rather than the function existing in a separate space. We can see the mathematical space as arising from the number systems that are being used to express the mathematical objects that are considered to exist in that mathematical space.

Of course this is not so obvious if we are just conceiving *sets*, and building structures based on that concept. But regardless of that, if a set consists of distinct objects then we *are* dealing with *numbers*.

There is no *scientific* reason why we should not also be able to take this approach when it comes to the Euclidean space of our everyday material world. We use number systems to make measurements of phenomena that pertain to this space. But in general, we also consider the space as if it exists independently of the phenomena that we may measure, when we choose to do so. The same thing applies to to the "dimension" we call time.

This view then easily translates into the psychological view that the object *spacetime* as it appears in the theory of relativity is also somehow an object that is distinct from the number systems and their functional relations that are used to theoretically embody the nature of all the objects it entails - namely, space, time, energy, mass, and gravity (or spacetime curvature).

The actuality is that this *is* just a *psychological* view, arising by extension from the way we commonly view everyday space and time as being independent of our knowing of it.

Our scientific theories for describing phenomena are constructs that have arisen out of the intelligence we are being, as it arises through brain function. Our assumption that they are about something else, something other than the intelligence we are being, just that - an assumption.

The truth is that there is nothing that we can know, that we can show is independent of our knowing of it. And therefore, nothing is independent of the way in which we *come to know*. It comes down to neuroscience. From the position of neuroscience, everything we know is also a construct of brain function, even though we don't know how that comes to be the case.

And that is why as we are exploring it here, we defined *genuinely objective* as independent of *individual* brain function, or *network consensus*, but not *independent of brain function*.

Essentially, we can expect that what we find to be genuinely objective in this way, is some part of the way in which brain function gives rise to our experience of the world. In other words, we cannot just expect the study of the brain *as an individual organ*, to throw sufficient light on this. We have to marry what we know to be genuinely objective about natural phenomena in general, with what we know about brain function, in such a way that we come to better knowledge not just of natural phenomena and something separate from us, but also to know it as part of the *principle of the brain*. That is, the principal through which we come to be this intelligence and experience of being, in the first place.

Part 3
Iterated Structure

13: Some Botanical Sketches

T hese "botanical sketches" might, from the look of them, have appeared in a book on plant life. But this isn't a book on plant life.

They weren't created using a human artist working freehand, or through computer art. They *are* natural objects. They are natural, mathematical objects. That is to say, they are literally, graphical representations of mathematical objects created through a simple process involving just a few simple rules.

A little later we'll look at how they are produced. They haven't been "doctored" or "improved" in any way to make them look more like a botanical sketch. What we are looking at is a "raw" representation of a *set*, a mapping onto the page of a set of complex numbers on the complex plane. Each "sketch" is a graph of a set of points that are a subset of the complex plane itself. They are arrived at through an IIP of a kind that might also be called an iterated function system or IFS.

To our eye and brain, these images are easily recognisable as being "plant-like", and with some "hand sketching" type features. Not only that, but we tend to see them as sketches of plants where the plant itself exists in 3 dimensions. In actuality, they are just 2 dimensional pictures of 2-dimensional mathematical sets.

So how is it they look so plant-like? The psychological answer is because our brain function is already "wired" to recognise and comprehend plant-forms, when we encounter them. What we are experiencing is an effect of neuroplasticity in brain function, an effect that psychologists call *pareidolia*, our brain interpreting incoming information in a way that makes sense according to already established patterns of cognition and recognition.

Like our faculties and features in general, pareidolia exists in the way our mind works, because of the past. Some would say it equates in some way to a survival advantage for our genetic form.

We all tend to see the world through a filter of already-existing comprehension. A filter created by our own brain and mind. A filter based on what we have already learned, and already understand, from the past. So pareidolia isn't just something that happens in our visual sense. It happens in our other senses too, and it happens in our ability to comprehend, and cognise. Our minds are always susceptible to *cognitive pareidolia*, because of the way our brain is already working.

Even though adaptation is a principle in evolution, it is enabled by nature not being in a state of balanced, unchanging "perfection". Nature - which includes our own brain - is dynamic, and makes "mistakes". Nature as a whole, as a system, is in a condition of perpetual change, somewhere between chaos and stability. That's how evolution can happen in nature in the first place. So even our comprehension, our cognitive ability, makes "mistakes", and our scientific knowledge of our world isn't entirely stable, but is in a state of evolution.

We, as beings, are not separate from nature, looking at nature, but are part of nature, out of which we spring. The challenge for us is to understand what nature *is*. Naive realism is the view that nature is the world as it appears to us, as sentient beings, through the principle of brain function.

The evolution of science since the 17th century Scientific Revolution, has progressively undermined this view. But it is still with us, despite all our scientific knowledge, including the knowledge of relativity, and quantum theory. The final frontier, of course, is the brain, and coming to terms with the fact that there is nothing we know or understand in science, and nothing we experience, that is not a construct a brain function.

"Nature-Like" Infinite Iterations

Mathematically, iteration may be best approached to begin with, in a simple, rather than complex system. Nature, however, exhibits all kinds of structures, from the simple, to the hypercomplex, our own brain being the prime example of the latter. And most of what goes on in nature, as far as we can now see, is non-linear, that is, it emerges from chaos.

Complexity in nature as a whole, and in specific examples such as ecology, involves very large numbers of interacting causes. It is somewhat unfortunate then, that the word *complex* is also used in the terms *complex numbers*, and *complex plane*.

Compared to the complexity is found in nature in general, there is nothing very complex about complex numbers, in themselves, or the complex plane, in itself. The "complex" term in the latter merely refers of course, to the presence of imaginary numbers in the mathematical structures. It is this kind of "complex" structure that leads to the "botanical sketches" that we have illustrated and are now going to talk about.

Let's begin with first principles. The so-called complex plane is a 2-dimensional mathematical space, that consists of the complex numbers. A complex number has two parts, the "real part", which is a quantity such as can be encountered in the material world, and an "imaginary" part which is a quantity of "i", defined as

$$i = \sqrt{-1} \,,$$

which we never encounter directly, in the material world. The real part of a *complex number* is one of its coordinates on the complex plane, and its imaginary part is the other. The imaginary part is conventionally represented on the vertical axis, whilst the real part is represented on horizontal axis.

So a complex number is a point on the complex plane. We'll now create a mathematical object through iteration on the complex plane. We're going to use the iteration function

$$z \rightarrow \mathbf{M} \cdot z$$

Here, z is a point on the complex plane, anywhere except at the origin. $\mathbf{M}$ is a complex matrix:

$$\mathbf{M} = \begin{pmatrix} x & x \\ x & x \end{pmatrix}$$

where x is another point, anywhere on the plane except at the origin. $\mathbf{M}$ causes a *transformation* to z, which moves it to another point on the plane. On the first iteration $\mathbf{M}$ moves z to somewhere else on the complex plane, and then the next iteration moves the new point to another point, and so on. Each iteration produces a new point, and all these points constitute a set.

The path taken by z is very sensitive to x, and in general, as the iterations continue, z will either spiral inwards towards the origin, or spiral outwards to infinity, in logarithmic spirals, depending on the value of x. Only for small values of x, smaller than around $0.4 + 0.4i$ does z *not* spiral out to infinity. But we can keep z *bounded* by introducing a bias d to the real part of x, so that $\mathbf{M}$ becomes:

$$\mathbf{M}_d = \begin{pmatrix} x+d & x \\ x & x+d \end{pmatrix}$$

Then, for example, in one case, where

$$x = 0.5 + 0.5i$$

and

$$d = -1$$

the point z will move continually around four corners of perfect square, and never escape from this pattern.

Essentially, an iteration such as

$$z \rightarrow \mathbf{M} \cdot z$$

implies in its existence an infinity of iterations. So a question with any such *iteration function* is what is the mathematical object that arises as a set, or space, through infinite iterations of the function, if each iteration value is added to that space?

In the case of $\mathbf{M}$, the object depends on x, but will usually involve either a "fall" of some kind, from the starting point into the origin, through infinite spiralling, or an "escape to infinity" in which it spirals outwards through the plane infinitely. There are of course infinite different patterns of each behaviour, but a few examples of the path that z follows, are illustrated here, for a given starting point that is $z = 4 + 4i$. The point z doesn't follow the path line continuously, but with each iteration moves from one junction to the next.

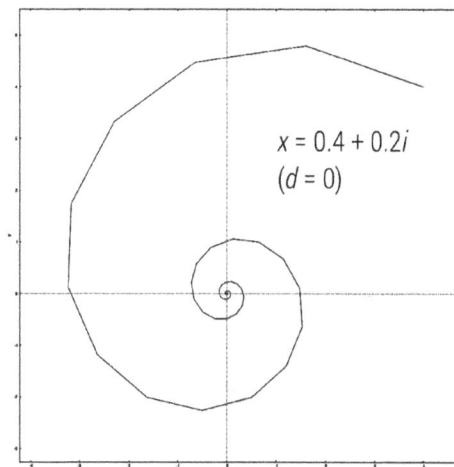

x = 0.4 + 0.2*i*
(*d* = 0)

When $x = 0.4 + 0.2i$ the point z spirals gracefully anticlockwise into the origin, as illustrated above. This general logarithmic spiral shape is one we might recognise in natural life-forms, perhaps an obvious example being in some shelled creatures. Most people will recognise it as one kind of "sea shell" or "snail shell" shape. Of course, the mathematics behind the formation of an actual sea or snail shell is far more complex, but somewhere, there will be an actual mathematical *relation* of some kind between this structure, and other logarithmic spirals in nature.

The scales of the illustrations here are left almost invisibly small, as they are not important for our purposes, but they range from just beyond z's starting value, $4 + 4i$, in the case of spiralling inwards towards the origin, and are in the order of 10^10 in the case of z escaping to infinity.

For the matrix $\mathbf{M}_d$ different behaviours emerge. For the value $x = 0.2 + 0.4i$ we find that for values of d greater than 0.2, the point z still spirals out to infinity. But as d reduces and approaches 0.2, the escape to infinity becomes far less rapid. The path of z begins to cross over itself, and becomes more angular. We can see this in the illustration of the path when $d = 0.9$ to 0.21, below.

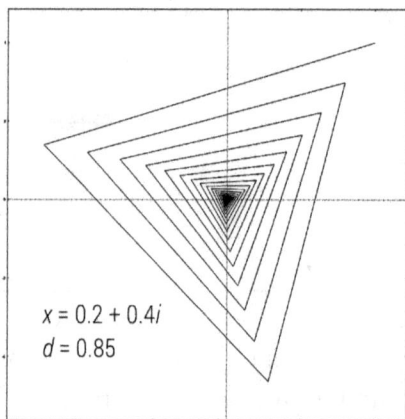

$x = 0.2 + 0.4i$
$d = 0.85$

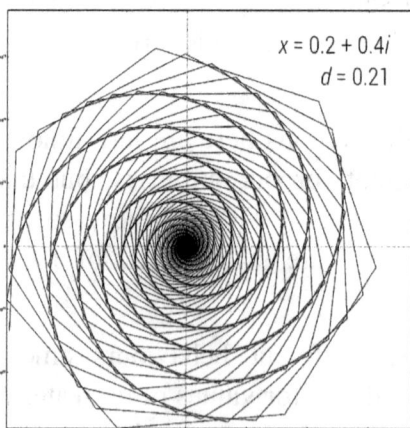

$x = 0.2 + 0.4i$
$d = 0.21$

When d finally reaches 0.2 then z becomes trapped in a *limit circle* around the origin. It moves forever to new positions on a circle around the origin whose radius is always the distance from the origin of the original starting point for z.

As d continues to reduce, z spirals inwards, but when $d = -1$ is reached, z circulates trapped again, neither escaping to infinity, nor spiralling inwards. The limit circle for $d = -1$ is illustrated below.

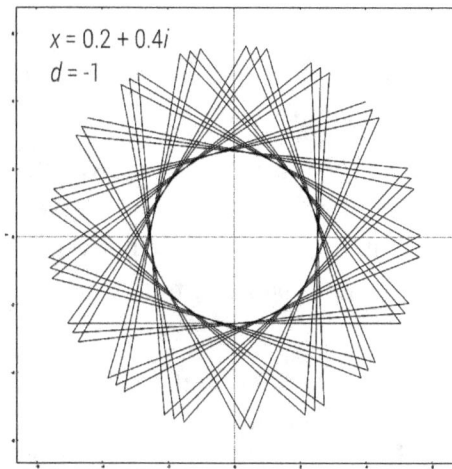

$x = 0.2 + 0.4i$
$d = -1$

So for this particular value of x in the matrix M_d, between the values of d at 0.2 and -1, z spirals to the origin, but outside this range of d, beyond the two *limit circles*, it escapes to infinity.

The relation of the z's path to M_d follows definite patterns but is nonlinear. Between paths of escape to infinity and spiralling to the origin some other limit circles occur, at the following values:

Re(x)	Im(x)	d	d
0.2	0.4	-1	0.2
0.4	0.4	-1.4	

Re(x)	Im(x)	d	d
0.8	0.4	-1	-2.2
1.6	0.4	-3.8	
3.2	0.4	-7	

We can see that when z is caught in the limit circle illustrated above, where d = -1, its path is one of triangular precession. This trend continues as d increases towards zero, but the rate of precession of the triangles relative to the iterations, changes.

The progression of behaviour as d is increased from -0.99 to -0.85 is illustrated in the pictures on the next page.

When d reaches minus 0.9 the path stops crossing itself, and in the third illustration the orbit is almost periodic. We can see that such iterative objects exhibit transformations between, and mixtures of, certain "archetypal" features, namely, multi-armed spirals, the circle, polygons, and "stars".

All of these arise from the iterative transformation of the position of a "point" through a "space", which is the *orbit* of the point. When the orbit is perfectly periodic then the value of the point keeps moving from one value to another, over a finite set of values that form the vertices of a closed polygon. None of the polygonal orbits illustrated above are perfectly periodic.

One might think that the circle is a periodic orbit, as it would be if this were the orbit of a material object moving through physical space. However, this is not as straightforward as it might seem. Of course, if we consider the circle as a polygon of infinite vertices, and we consider the point as moving sequentially around the circle through the vertices, then we would have periodicity.

In actuality, though, this situation cannot arise from an ordinary iteration equation or iterated function system. Such systems, even when the iteration is infinite, cannot output the desired orbit unless we were somehow able to the lengths of constructing an iteration that uses infinitesimals.

It is possible for the orbital positions of the point to *tend to* a circle over infinite iterations. But the orbit is never periodic unless the same set of points in the space are repeatedly visited by the orbiting point. The machinery of iterated equations and IFSs do not suffice for this kind of structure.

In the examples illustrated the path of the point never quite closes the polygon, resulting in a rotation or *precession* of the orbit, that arises from the fact that the orbit is not perfectly periodic. It is easy to see how *multi-armed spirals* are formed from polygons that are not perfectly periodic, and have precessing orbits.

Below are some more examples showing how periodicity and behaviour that is not perfectly periodic creates the relations between processing polygons and multi-armed stars.

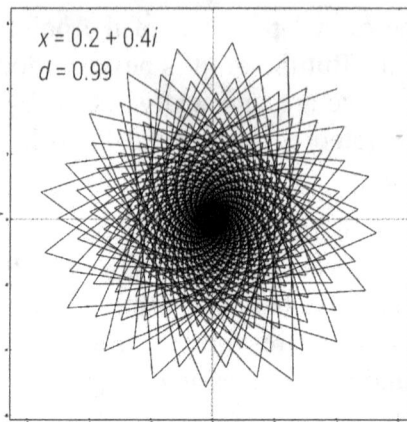

$x = 0.2 + 0.4i$
$d = 0.99$

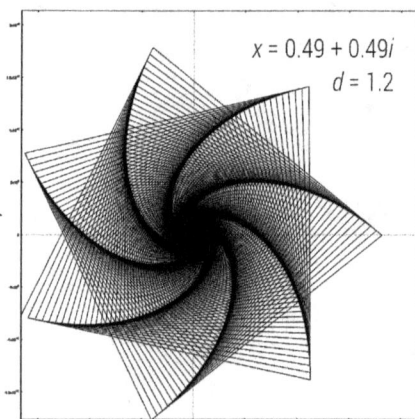

$x = 0.49 + 0.49i$
$d = 1.2$

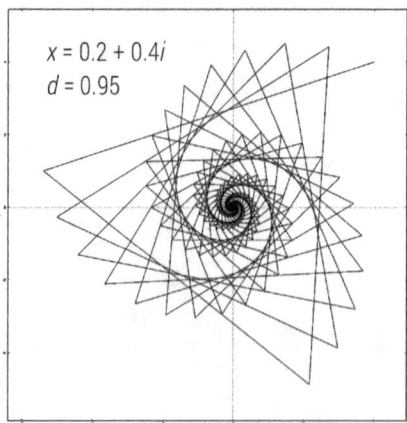

$x = 0.2 + 0.4i$
$d = 0.95$

Spaces Versus Paths

What we have been looking at here are pictorial representations or graphs, of finite parts of the mathematical object produced by the infinite iteration function:

$$z \rightarrow \mathbf{M}_d \cdot z$$

The matrix $\mathbf{M}_d$ depends on x and d, so the general form of this is an iteration function f, as:

$$z \rightarrow f(x,d,z)$$

Each of the parameters x, d, and z, are points in the complex plane, so what we are looking at are some exotic ways in which points on the complex plane can be related to each other simply through an iteration function.

The function creates a subset of the space of the complex plane through the process of iteration, and what we are seeing here is the *path through the plane* taken by the point z.

Rather than concentrating on the *path* taken by z through the complex plane, although interesting, we can create a "botanical sketch" from the *set of points* that z takes, generated by the iteration. To do this we introduce *scaling* and *translation* components into the function, and we need also to introduce a random *probability* element into it.

The iteration function we use is (with no d bias):

$$z \rightarrow \mathbf{M}(\mathbf{S}_n \cdot z) + c$$

That second matrix $\mathbf{S}_n$ is a scaling component that occurs at random on each iteration, with equal probability from three possibilities:

$$S_1 = \begin{pmatrix} D & 0 \\ 0 & D \end{pmatrix}$$

$$S_3 = \begin{pmatrix} 0.98 & 0 \\ 0 & 0.98 \end{pmatrix}$$

$$S_2 = \begin{pmatrix} 0.99 & 0 \\ 0 & 0.99 \end{pmatrix}$$

D is a "dissipation" component, a function of the number of iterations, that damps the process by approaching zero as the number of iterations increases. This is rather like the amplitude of swing of a simple pendulum, which is damped due to friction or dissipation of energy. Here, it is given the value

$$D = \frac{R - k}{R}$$

where R is some arbitrarily large number of iterations, and k is the iteration number of each cycle. The final parameter c is another point on the complex plane, this time much closer to the origin than the starting point, although it doesn't have to be. Here, it is $0.001 + 0.001i$. The equality of the real and imaginary parts simply means that the final object is upright. Changing the magnitude of the number will change the scale of the object. This is a "damped" *translation* parameter that just moves z on the complex plane by a small amount, with each iteration.

The iteration now produces the "botanical sketch" of the kind we saw earlier, at around 250,000 iterations.

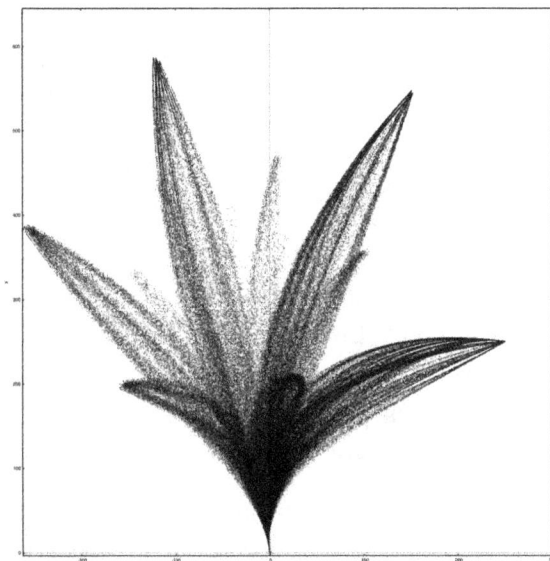

Essentially, the difference between this and the geometric paths we also saw, lies in the four additional principles being added into the iteration. These are scaling, translation, dissipation, and randomness.

The components of scaling and translation come under the umbrella of *transformation*, that takes place on a point when it is acted on by a matrix.

Dissipation and randomness are ubiquitous components in the complex causal networks of nature and life forms. Throwing these other components into the recipe of the iteration, so to speak, produces a different "kind" of object - in this simple case, the "botanical sketch".

Varying these parameters will change the object somewhat, but the basic "idea" of it, as generated by the iteration, if we can refer to it that way, remains the same. However, the system still depends on the matrix M_d and *its* demands for keeping the point z from escaping to infinity, or collapsing into the origin too fast. If there is just the right *small range* of conditions provided by the matrix to begin with, then when the new transformation, dissipation, and randomness elements are introduced, the "sketches" can arise.

In doing so, we move from the highly ordered world of stark mathematical, geometric form produced by $\mathbf{M}_d$, the stars and polygons and spirals, to another kind of order, more like the kind our brain recognises in life forms.

Life forms are products of interactive complex causal networks and actually, we might observe, the principle of iteration. The principle of iteration is in life forms because they all come out of the principle of recurrence, through which genetics is able to evolve forms through natural selection and other evolutionary processes.

In life forms, there is mathematical form and structure in biological mechanisms that become capable of producing the kind of order we recognise in life forms, also because of the presence of the principle of random probability, and dissipation. But there is a one big difference to how it is done in life forms, and how we did it here. And that is the presence of the *complex, causal network* (complex in the sense of vast numbers of interacting parts).

We have just used an iterated function system, which in this case happens to be a very *small* and *simple* network of functions, that we could depict as a single system "block" something like this, in which the inner vertical block signifies equal probabilities of each $\mathbf{S}$ matrix occurring on any iteration:

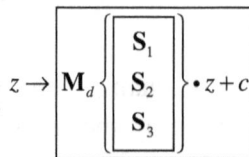

$$z \to \mathbf{M}_d \left\{ \begin{array}{c} \mathbf{S}_1 \\ \mathbf{S}_2 \\ \mathbf{S}_3 \end{array} \right\} \cdot z + c$$

Iterations in nature can happen through much larger and more complex systems, in which there are iterations within iterations, such as the iterations of DNA production within the iterations of whole-organism lifetimes.

In the case of life forms, iterations are taking place *in time*. In the context of mathematical form, the only way in which the iterations we have been looking at, depicted in the pictures, take place in time, is because to produce the pictures they take place through the cycles of the computer program that produces them, or the the cycles of the computer processor.

Mathematically, the iterations of our function system are not themselves a function of time. The entire possible output of the system is instantaneously implied in the function system itself. It's what the function system *stands for*. By writing

$$z \rightarrow \mathbf{M}_d \left\{ \begin{matrix} \mathbf{S}_1 \\ \mathbf{S}_2 \\ \mathbf{S}_3 \end{matrix} \right\} \bullet z + c$$

and defining the matrices in the way we have, we are in effect symbolising the "botanical sketch" itself, as a symbolic structure. Because there is a probability element in the function system, there is a range of outputs that the system can produce, or that the box above symbolises. Until the iteration actually takes place, the result is non-deterministic. The object that the function system stands for, until the iteration takes place, is a probabilistic superposition of all possible outputs.

The way this particular system is set up, the random variations between one possible output and another, are small. But nonetheless, if you run the iteration on a computer, each time you will get a slightly different result. If we replace the probability with a certainty for one of the **S** matrices then no "botanical sketch" results. *Uncertainty is part of the "game"*.

When we looked at the *paths* taken by the point z in the first set of iterations, we saw "meaningful" geometric patterns. The *path* of z in the case of the "botanical sketch" is rather different in appearance. It's illustrated here for the first 4000 iterations:

"botanic sketch" as the path of *z*

The *path* of the point *z* shows how its position changes with each iteration, but for the "sketch" it would seem the path is not so important as the *set of points* that results. However, the set of points *only arises as a consequence of the path*. The path is pre-requisite. The general principle is that an IIP leads to the path, which leads to the form.

Of course, in practice we cannot iterate infinitely, but the kinds of iteration systems that we have been looking at don't demand this. Rather, the resulting space is already contained within an outer limit, so increased iteration just fills in more and more of the detail of the object.

So an iteration function or iterated function system of this kind, is such that we could symbolise it :

$$\boxed{\{f\}(\mathbb{S})} \rightarrow \Omega \subset \mathbb{S}$$

where the box on the left represents the system containing the set of iteration functions $\{f\}$ that act on the space $\mathbb{S}$, to produce the object Ω (the botanical sketch) over infinite iterations. The object Ω produced is a subspace of the space $\mathbb{S}$, which in this case is the complex plane.

Now an important point is this: To get from Ω to a "botanical sketch" actually requires something else. First, it requires a mapping from Ω which is a mathematical subspace of the complex plane, with complex numbers, to a space of real numbers in a space like the space on the page. And it requires some kind of printing system that can map the set of real points onto the page. That's easy enough.

But this still doesn't explain literally why it looks like a sketch of a plant. That requires the intelligence of the reader who is capable of *recognising* the image. Because it's *not* a picture of a plant.

What we are encountering here, is what in psychology is called *pareidolia*. It is the brain's way of interpreting what we see according to what it already knows. And this will happen with very incomplete information. So what we are looking at is a matter of both mathematical objects, and brain function.

The last illustration shows the "botanical sketch" set of points, with the path of the points generated from the iteration, *traced*. There is a world of difference between the *path* the iteration values take, and the set of points that they constitute.

Infinite Iteration Processes

We have been considering things as "paths", "points", "sets" and "spaces'. There is, however, another way of looking at this. The iteration function system that we used:

$$z \rightarrow \mathbf{M}_d \left\{ \begin{matrix} \mathbf{S}_1 \\ \mathbf{S}_2 \\ \mathbf{S}_3 \end{matrix} \right\} \bullet z + c$$

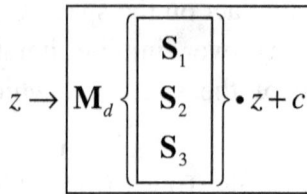

is itself an IIP:

$$\circlearrowleft (\infty) \{Z\}$$

where Z is itself the *structure* of processes that we defined in the box. What is the generic nature of that structure? It has the following features:

$$Z \equiv z \longleftrightarrow \left\{ \begin{matrix} \text{Orbital Balance} \\ \text{Scaling} \\ \text{Translation} \\ \text{Dissipation} \\ \text{Randomness} \end{matrix} \right.$$

in which all these features are interconnected in the network of objects that constitute the IIP generating the image.

Orbital Balance: This is the factor that prevents the orbit from either escaping to infinity or falling into the origin, too fast.

Scaling: This factor that determines both the size of the "sketch", and its orientation.

Translation: This factor "misaligns" points in the orbit from where they otherwise would have been. It is also affected by dissipation.

Dissipation: This factor causes the orbit to decay. It is intrinsic to Orbital Balance, but also causes translation.

Randomness: This factor randomly applies conditions determined by the above, drawn from three possibilities.

There is little point in trying to further unpack the nature of Z, although of course we could do so, if we wished. We are looking here at an IIP that produces the "sketches" by being based on "just the right mixture" of these features being thrown into the complex causal network of factors on which the IIP does its work.

We began with the basic ability of an iteration function to produce *iteration patterns*, such as the precessing polygons, stars and spirals that we saw above, and then we "disrupted" the process by adding in the other factors. This is not the work of "computer graphics" but it is essentially a mix of art and mathematics.

It is a relatively simple way of showing the coming together of certain mathematical features connected in a *structure*, to which is applied to the principle of iteration. Iterated function systems, or IFSs, in general use this principle, and the IIP we used here could be considered as such. It results in a specific kind of "meaning"- the plant-like "sketches" - when it is encountered in human intelligence, provided the *process* is *represented* in a suitable way - in this case, as an image.

Note that the "sketches" are *not* what would generally be considered *fractal*, although they do each exhibit parts that are similar and at different scales. We can easily see that we could extend this principle into wider structures of IIPs where the wider structures themselves, are IIPs. Furthermore, there is no reason why some IIPs in such a structure should not be infinite fractal generating IFSs.

We have described the mathematical components of the IIP we have been using as:

- Orbital Balance
- Scaling
- Translation
- Dissipation
- Randomness

which is their specific mathematical roles. Similar features are found in iterated function systems generally.

However, there are other ways to consider what is going on. It is possible to consider this entire process we have been using as one with the following structure of relations between the following objects:

ITERATION

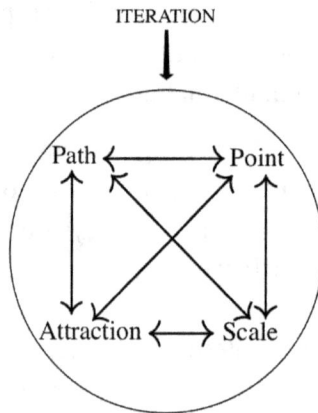

In other words, the machinery of the IIP has four salient objects that are *point*, and *path* (of the point), *attraction*, and *scale*. Let's discuss the principles in depth.

The Point

Can there be a "point" that is not a point on the plane, or in a system of coordinates?

The mathematical "point" object is a number, or set of numbers. The fact that we relate it to a system of coordinates or set of points in which it is a subset, is a tool of understanding that we are using, and not an implicit feature of the way the number behaves under iteration. In the conventional view, however, the point appears not to exist without the coordinate system, or complex plane, or set of points of which it is a subset.

The structure we have already conceived includes both objects, the *point* and the *plane*. We conceive the point as existing *relative to* the plane, or as a *subset* of the plane. We don't normally consider the plane's "set of points" to be dependent on the iterating point, but in fact it is, just as the plane is dependent on any object that consists of points or complex numbers, that is ordinarily considered as existing "in the plane". We tend not to perceive the structures we are looking at because we don't think in terms of structures detached from the meanings for us that they give rise to.

The plane itself, when considered as a "set of points", each of which is a number, is a construct that is part of the total way in which we conceive what we are looking at and the behaviour we see. However, we don't see anything like the complex plane itself in natural phenomena, other than in the natural phenomena of our mind. What we do see is continuum quantities and their relations that we can model using the construct of the plane and its numbers. Above, we interpreted these continua quantity relations in nature as instances of the continuum ♣.

Where they occur in natural phenomena there is then a structure that includes this object, that gives the continuum quantity in question, such as electromagnetic radiation for example, its particular *qualitas*.

The complex plane - the theoretical *continuum* we are using in mathematical modelling, that corresponds to this, is the object:

$$\circlearrowleft (\infty) \{ \clubsuit \}$$

We commonly conceive this abstractly, mathematically, as the *complex plane* continuum. At the same time, we presume numbers to be capable of being put together in a structure that is the same continuum object, and have gone to great lengths to build abstract

Page 137

theories of how this can be done, using ideas such as hyperreal numbers and so on.

In terms of objects and structures whenever we assign a complex *number P* to a position on the *continuum* of the plane, we instantiate the object:

$$\circlearrowleft (P) \{\clubsuit\}$$

where now $\circlearrowleft (P)$ is the IIP that produces the complex number P that is *applied* to that particular *quantity* on the complex continuum $\clubsuit$, rather than being the same thing as that quantity.

We think "A quantity is a quantity and is a number". But we are mistaken in that conception. We build structures upon axioms that we call "proofs", just as Euclid did, to this effect, and become satisfied that we have found absolute truths. Today, we are not where the ancient Greeks were, we are in fact in the wider picture ready to explore where we have never been before, but then, our conviction that mathematical proofs are genuinely objective, in a timeless way, and not just a matter of network consensus, closes our comprehension.

The object we call an iterating "point" is actually the state of the IIP producing the point at any iteration number n where n always progresses through the natural numbers N. This iterating "point" *exists as* the structure:

$$\circlearrowleft (P) \{\clubsuit\} \longleftrightarrow \circlearrowleft (\infty) \{N\}$$

where P on the left is the complex number assigned to its position on the complex plane continuum, and the IIP on the right is the IIP for the natural numbers, and the *relation* $\longleftrightarrow$ is the IIP that creates the path of the point, that produces the image object, the "botanical sketch".

Things get a little more involved when we apply this to an object such as a Julia set or the Mandelbrot set, because the IIP in that relation follows a different principle altogether. There, it is not the path of the point that produces the final object such as the Mandelbrot set, but we shall look at that later.

We have simply *not written in* that IIP *specifically*, we have just left in the arrows showing the *relation* between the other IIPs. This is because we already *know* this relation is the IIP that produces the image object. We can still do the same thing with an object such as the Mandelbrot set.

The "set of points" that are complex numbers, that we ordinarily say *constitutes* the complex plane, cannot be created from a single IIP. We saw above how there is no such thing as an IIP for creating consecutive real numbers, and therefore the same is true for the creation of the "set of points" that are said to constitute the complex plane.

Complex plane "points" are a construct of convenience, in the same way as is the notion of say, the infinity of "points" on a "real" line. These constructs break down when they are pushed too far, precisely because a "point" constitutes a *number*, and the *processes* that real numbers *exist as*, or are parts of, cannot be genuinely made into a structure that is a continuum. In other words, real (and hence complex) numbers can be used to express continuum quantities, but not the quantity continuum itself.

The iterating "point" we are talking about here, as a structure, is simply the IIP:

$$\circlearrowleft (P) \{♣\} \longleftrightarrow \longrightarrow \circlearrowleft (\infty) \{N\}$$

This expresses how the iteration (contained in the relation arrows) links the IIP of the natural numbers (on the right) to the IIP that creates the complex number values P for the orbiting "point" z, that are applied to the complex continuum ♣.

The Path

The complex plane continuum - our theoretical construct that stands for the natural complex quantity continuum ♣ - has the structure:

$$\mho \left(\infty \right) \left\{ \left[\mho \left(\infty \right) \left\{ \Upsilon \right\} \right] \right\} \left\{ \Omega \right\}$$

which symbolises the IIP of the imaginary continuum Υ infinitely iterated over the IIP of the real continuum Ω. We could equally well have said:

$$\mho \left(\infty \right) \left\{ \left[\mho \left(\infty \right) \left\{ \Omega \right\} \right] \right\} \left\{ \Upsilon \right\}$$

which is expressing it the other way around, as the IIP of the real continuum infinitely iterated over the IIP of the imaginary continuum.

So the *path* of the iterating point on the plane is a structure made from the two component objects, one for the point, and one for the plane:

$$\mho \left(P \right) \left\{ ♣ \right\} \longleftrightarrow \mho \left(\infty \right) \left\{ N \right\}$$

and

$$\mho \left(\infty \right) \left\{ \left[\mho \left(\infty \right) \left\{ \Upsilon \right\} \right] \right\} \left\{ \Omega \right\}$$

We can easily write these structures as objects in a more condensed way by saying:

$$P \equiv \left[\mho \left(P \right) \left\{ ♣ \right\} \longleftrightarrow \mho \left(\infty \right) \left\{ N \right\} \right]$$

and

$$C \equiv \left[\mho \left(\infty \right) \left\{ \left[\mho \left(\infty \right) \left\{ \Upsilon \right\} \right] \right\} \left\{ \Omega \right\} \right]$$

so that we can more briefly just say:

$$P \longleftrightarrow C$$

What are we looking at here? P is the changing complex coordinates number z that constitutes the *path*, and C is the complex plane as a continuum, and the relation $\longleftrightarrow$ is the IIP we know is:

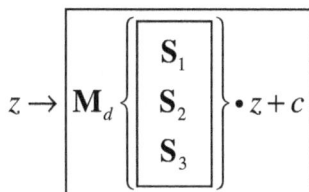

$$z \rightarrow \mathbf{M}_d \left\{ \begin{matrix} \mathbf{S}_1 \\ \mathbf{S}_2 \\ \mathbf{S}_3 \end{matrix} \right\} \bullet z + c$$

P and C are each infinite *iteration processes*. P is one of numbers, but C is the IIP of a continuum. To summarise this again, P's structure is the IIP creating an infinity of distinct objects called *natural numbers*, each causing the IIP to generate a complex number. This is then related to the *continuum* C that we are calling the *complex plane*, by being expressed as a number-defined "position" on it.

So we can see that even though the IIP for z, in the box, that we use to create the "sketch", is itself based on numbers, points, and the construct of a complex plane that consists of points, the *path itself* has a generic structure that is an IIP in its own right.

We are saying that the path

$$P \longleftrightarrow C$$

can be considered as an object in its own right, as the path P of complex values related to the natural numbers (as the iteration numbers), which is then applied to the complex continuum C. The precise *way* in which the path arises in relation to the natural numbers is of course a matter of the whatever IIP is locked up in that relation $\longleftrightarrow$ in the structure. Always this IIP is directly related to the IIP of the natural numbers, because it is the specific iteration function or system for iterating z. It doesn't necessarily have to be one that creates "botanical sketches". It can be far more interesting, or far less interesting.

We are looking here at a generic form of structure. The *path* and the *continuum* are distinct objects. In nature, we don't see such "botanical sketches". We see actual plants, and many other things. The processes that produce them are not in general structures of the kind we have just looked at, which is the process wrapped up in the relation arrows in the structure:

$$P \longleftrightarrow C$$

That was a mathematical structure:

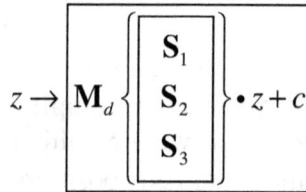

$$z \rightarrow \mathbf{M}_d \left\{ \begin{matrix} \mathbf{S}_1 \\ \mathbf{S}_2 \\ \mathbf{S}_3 \end{matrix} \right\} \cdot z + c$$

that is relatively simple to the structures we find in nature in general. In general, natural forms and processes, especially in living organisms, and biological systems, can be expected to be large networks (structures) of objects (processes) that are far reaching into the structure of nature as a whole. This is not least because what we encounter as actual plant forms, for example, are not encountered in the first place, except through the structure of brain function.

Nevertheless, there *are* structures in natural phenomena, many of which we find in classical physics, that can be mathematically encapsulated in a simple equation, or a few simple equations. This brings us to the point that *simplicity* in nature is deceiving. These simple equations lead us to believe in a simple nature, because we tacitly think that the *numbers* that such equations represent relations between, are the same thing as, or perfectly represent, the natural

Page 142

quantities they stand for. The truth is that they do not. The actual quantities in natural phenomena, that we encounter, are the result of structures of complexity that we currently have no conception of.

What we have done, in the equations of classical physics, is to build mathematical structures representing the relations and behaviour of the natural quantities we encounter, that are completely dependent upon number processes, and that only make sense to us in terms of concepts like "mass", "energy", and "length", which are not themselves otherwise understood either as numbers or as structures other than of each other.

The structures of these natural quantities that we encounter, the structures that give each its qualitas in the way that we encounter it, cannot be separated from the means by which we encounter them, which is the structure of brain function. This is both a structure of vast complexity, and a structure of processes.

Conversely, complexity as infinite convolution can arise from seemingly great simplicity, as we see in objects such as the Mandelbrot set, where an IIP is involved. We will look at this later.

Attraction

The principle of "attraction" and its opposite shows up in complex dynamics in *attractors*, and their opposite, *repellers*.

In general, iterations on the complex plane create orbits that tend to an attractor, or remain stable, or remain chaotic, or *escape to infinity* (which may also be considered to be an attractor). Often, the fate of an orbiting value of z uncertain, because it is only possible in practice to iterate finitely. In other cases it is possible to show that z is definitely falling into an attractor, or escaping to infinity. In parallel to the physics of orbits, we can still regard a stable orbit around an attractor, as a "falling", if we want to, just as a means of description.

However, unlike actual satellites and their gravitational attractors, it is not the case that attractors and repellers on the complex plane exist on the complex plane in a way that is not dependent on the orbiting object - point produced by the IIP. In fact, the role of the complex plane whether we consider it geometrically or not, is effectively only as a useful theoretical theatre through which we can comprehend some meaning in the behaviour of these numbers - or to be more specific, the behaviour of the *relations* between the number objects that z becomes.

Speaking of attractors and repellers as though they are things in their own right that exist on the plane, can be useful. But actually, the only thing that exists in this scenario, is the relations between numbers that z becomes.

In the "botanical sketch" IIP we introduced a factor specifically to influence the orbit in order to create the "sketch", rather than allowing the orbit to escape rapidly to infinity, or to fall into the origin.

There is a broader way of looking at all this. It is possible to categorise the dynamics using just the concept of *attraction*. We can do this by regarding repulsion as a negative attraction. Also, we can regard attraction to an attractor as a phenomenon that manifests in two ways:

o The first way happens where the orbit is attracted towards a position on the complex plane.

o The second way happens where the orbit *escapes to infinity*. Attractors appear in the complex plane *as if* attraction to a position on the complex plane is not a consequence of the iteration, when in fact it is. Similarly, infinity itself is an attractor, but infinity is not a part of the complex plane.

In this way, there is attraction to a finite position, and attraction to infinity. There may be many different kinds of attractors on the complex plane, in a complex system, including those that repel, but they are all fundamentally of one class, because they all attract the orbit towards a position in the complex plane, or an area of the complex plane, that is finite, and can be described by a number.

In contrast, attraction to infinity, because it cannot be recognised as attraction towards any particular part of the complex plane, and most certainly cannot be identified by any number or numbers considered to be part of the complex plane, is an *attractor* of a different kind.

In fact, both attraction towards attractors in the plane, and an escape to infinity, are both inherent in the nature of the iteration itself, rather than something that exists on the complex plane independently of the iteration. Specifically, infinity does not exist anywhere *on* the complex plane, or as part of it, when the complex plane is considered as a set of points that are numbers. Infinity is not a number.

However, when we recognise that infinity, as it appears in the phenomena we can contemplate, is always encapsulated or represented in a *process*, then an alternative picture emerges.

Now we can begin to see the structure. We can see that a path $P \longleftrightarrow C$ as determined by the IIP producing it, can in general be related to either or both a finite attractor object [A], and/or infinity. Finite attractors and infinity, are both IIPs. So what we are looking at as the *path* of z, is the IIP for the path, that may be attracted to a finite number's IIP, or the IIP for a continuum (which being an *infinite* iteration process, is the structure of an *infinite* continuum).

When the IIP of the path "escapes to infinity", what that means in simple terms is that the magnitude of z is continually increasing with each iteration. There is no such thing as a *value* of z that is "infinity". So as long as we are looking at the relation between z as a number, and the iteration number n:

$$z \longleftrightarrow n$$

then that in itself is not the actual escape to infinity, but rather, the process of z as a number, as it were, *in the process of* escaping to infinity. The number z has no fate in this iteration other than to remain a finite number. In iterations such as those used for the Julia sets or the Mandelbrot set, the term "escape to infinity" refers to the fact that the orbit of z as n increases does not remain bound to any attractor on the complex plane, in other words, it *escapes* from orbiting attractors on the complex plane. The number z continues to change with the iterations, and just as the iterations continue infinitely to follow the sequence of the natural numbers, which is infinite, so z itself never "arrives at" the infinity towards which it is now being attracted.

The situation we are looking at is actually one of infinite *processes*. Even an iteration for which z *does not escape* is an infinite process, whether or not the orbit is periodic. In fact, even if the orbit of z leads to a repeating number for z, as it does for example when the iteration is

$$z \longrightarrow z^2$$

and the initial value is $z_0, = 1$, then there is still an infinite process that is the infinite instantiation of the number z (in this case 1).

The essential difference between an IIP that is an orbit that does not "escape to infinity", and one that *does*, is a question of *scale*. We'll come to that in the next section.

The infinite IIP for the complex continuum C is the structure we already saw:

$$C \equiv [\text{Ծ} (\infty) \{[\text{Ծ} (\infty) \{\text{Ծ}\}]\} \{\Omega\}]$$

which is an IIP for a continuum. So the relation of the path to these two kinds of attractor is:

$$[A] \longleftrightarrow [P \longleftrightarrow C] \longleftrightarrow [C]$$

where the object [C] on the right is the complex continuum as a numberless IIP, and the other two objects, the path and the finite attractor, are IIPs for numbers.

As we said, the path $P \longleftrightarrow C$ is not really something "attracted to" attractors that pre-exist on the complex plane. Rather, the path, in effect, *creates its own attractors*. We could symbolise this neatly as:

$$[A] \leftarrow [P \longleftrightarrow C] \rightarrow [C]$$

Here we have now replaced those relation arrows either side of the path object, with single directional arrows. This is a suitable thing to do, because it indicates that the attractors are a consequence of the IIP that produces the iteration of z.

According to the premises of Object Theory (Premise 6) a *relation* (that we symbolise here as a double arrow) can be considered as an object in its own right, so if we apply that principle then the structure is:

$$[A] \leftarrow\rightarrow [A] \leftarrow \overset{\leftarrow}{-} \rightarrow [P \leftarrow\rightarrow C] \leftarrow \overset{\rightarrow}{-} \rightarrow [C] \leftarrow\rightarrow [C]$$

where the single directional arrows indicate the nature of the relation in each case. They in effect state that the path object $[P \leftarrow\rightarrow C]$ *produces* the attractors $[A]$ and $[C]$, which we already *know* is the case.

The structures $[A]\leftarrow\rightarrow[A]$ and $[C]\leftarrow\rightarrow[C]$ each now each appear as two distinct instantiations of the same object. As far as the $[A]$ is concerned this is just two instantiations of the same position on the complex plane, which is a theoretical construct that here, does not mean anything other then $[A]$.

We can go on replacing the relations arrows in the same way, with further instantiations, *ad infinitum*. This kind of recursion starts to become meaningful if the objects are not identical, but here, they are identical. A further explanation of this can be found in the explanation of Premise 6 at the beginning of this book. Essentially, A itself is an IIP that can be considered both as the IIP of its own infinite instantiations that are otherwise unrelated, and the more familiar IIP that produces the number that it is.

As far as the $[C]$ is concerned, we end up with infinite instances of the IIP for the complex continuum C, with never any relation between them other than that of the IIP itself. This also can be considered to be simply the same as the IIP of the continuum C. If we have $C \leftarrow\rightarrow C$ and the relations $\leftarrow\rightarrow$ are always the same every time we replace them with C, then we end up with C because there are no distinct objects in the structure, only distinct instances of the same object. Symbolically:

$$\circlearrowleft (\infty) \{C \leftarrow \frac{C}{-} \rightarrow C\} \equiv C$$

Scale

Many networks have the property of being *scale-free*, meaning that the distribution of connections approximately follows something such as a power law distribution, or a lognormal distribution. Such networks are widely found in nature. This characteristic results in *hubs*, that are nodes in the network that have a much larger number of connections than most other nodes. They are *attractors*, in their own way, a way in which generally, the bigger the hub, the greater the attraction.

We see this in the World Wide Web, in the domination of major hubs such as Google, Facebook, and so on, and even in the correlation between the connectivity of websites and their market value.

There is preferential tendency in the dynamics of networks for new nodes to connect to hubs, where overall, the larger the hub, the greater the probability of connection. The bigger the hub, the more it tends to get bigger. This is a now well-known principle in scale-free networks.

All of this is connected to the principle of *scale*. Networks that grow by evolving, such as natural networks, but also including human made networks such as the World Wide Web, are very often found to exhibit *scale free distribution*, or something quite like it (the term refers to the probability distribution of finding a given connectivity between any node, and the rest of the network). In a *scale free* network the structure of the network, in terms of the distribution of direct connections between nodes - otherwise known as its *topology* - is independent of the size of the network. As the network grows, its topology remains much the same.

Basically, this also means that for a very large and complex network, the topological appearance of the network - the characteristics of its configuration of connections - looks much the same as we "zoom in" to the network to see more details of its connectivity.

In contrast, a very large network with random distribution may appear dense when viewed from afar, but as we zoom into it, it looks more and more sparse. A network with scale free distribution appears more full of features, as it were, because parts of it are densely connected, and other parts are more sparsely connected.

This is reflected even when we zoom in to such a degree that we can see the individual nodes, in that there are a large number of nodes with relatively few connections into the rest of the network, whilst other nodes have more connections, and so on, up to a very small number of very large nodes, with the largest number of connections.

Scale free distribution is by no means true of all natural networks, however. For example, it is not true of the network of atoms in a crystal. Perhaps most notably, it may be that it is not true of the human brain, which is the most complex network known. At the time of writing this, however, there is not a widespread consensus on the network topology of the human brain. Nonetheless, it *may* be true of natural neural networks in general, and at the current time we *do know* that the neural network of the roundworm *c. elegans* does not appear to be scale free.

In all of this we can see a picture emerging in which we might speculate that there may perhaps be a way in which the manifestation of attractors in complex dynamics is linked to the topology of the *structure* of the IIPs that are in play. This is because a structure is essentially a network. We would be unlikely to begin to see this, though, until we had sufficient picture of all the IIPs in play, as it were.

Briefly, from what we have already said, it would be possible, in principle, to consider *any* object in *any* structure, as an IIP, or part of one. So given that anything that we might be considering is essentially a network of objects, or a *structure*, it may be, we might speculate, that the highest level representation of anything we might be considering, is as a structure of IIPs.

Part 4
Fractals

14: Fractal Structure

The Mandelbrot Set

Arguably the most famous fractal now is the Mandelbrot set. The Mandelbrot set arises indirectly from the iteration function

$$z \longrightarrow z^2 + c$$

or in our adopted notation:

$$\circlearrowleft (\infty) (z \longrightarrow z^2 + c)$$

where z is a complex number that changes with the iteration process, and c is another complex number that stays constant throughout the iteration process. However, the iteration process is applied to c values throughout the complex plane.

The iteration equation was originally investigated in relation to the infinity of Julia sets. The Mandelbrot set and the infinity of Julia sets are accordingly related.

We will be talking here about "filled" Julia sets, and only later will come to the former Julia sets, which are the boundaries of the "filled" sets.

In the first step in producing "filled" Julia sets, the c is held constant, and the iteration can be run for infinitely many different starting values (z_0) of z. For each run z either remains bounded within some boundary, or it escapes to infinity, such that as $n \rightarrow \infty$, $z \rightarrow \infty$. The second step is that any starting value (z_0) for which z does not escape to infinity is designated as a member of a "filled" Julia set for that value of c. This can be done for infinitely many values of c.

In the first step in producing the Mandelbrot set, the starting value is held at $z_0 = 0$ and the iteration can be run for infinitely many values of c. Again, for each run z either remains bounded within some boundary, or it escapes to infinity, such that:

$$\text{as } n \longrightarrow \infty,\ z \longrightarrow \infty$$

The second step is that any value of c for which z does not escape to infinity is designated as a member of the Mandelbrot set.

The boundary of the z_0 values between iterations that escape to infinity and those that do not, when $c = 0$, is a circle of radius 2 around the origin. This is the Julia set for $c = 0$.

The boundary for the Mandelbrot set is the infinitely long fractal boundary around the familiar Mandelbrot set object:

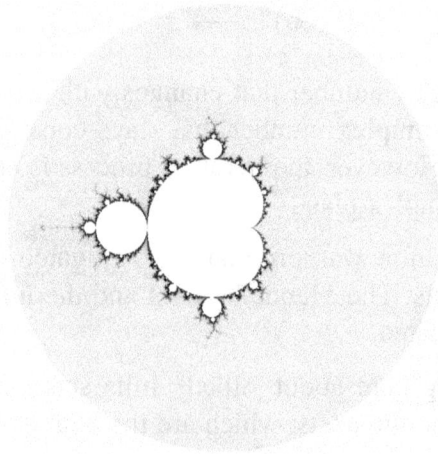

Julia sets for which c is a member of the Mandelbrot set, are connected, whilst those for which c is outside the Mandelbrot set are disconnected.

The boundaries are the parts of these objects where we find particularly beautiful and intellectually exhilarating fractal forms and behaviors. Today, in widely available computer generated interactive fractal imagery, this remarkable form of beauty is rendered by colouring schemes related to the *rate* of z's escape to infinity (in relation to iteration number), at any point in the image. For z_0 in a

Julia set, or for c in the Mandelbrot set, if there *is* an escape to infinity for that value, the *rate* of escape depends on z_0 or c respectively.

Inside the Mandelbrot set, where $c = 0$ then z remains zero (at the origin of the complex plane). For c $\neq$ 0 inside the set, z falls or cycles in oscillatory point orbits or lines towards an attractor, or multiple attractors other than the origin. The positions of attractors and values of z surrounding them, in iterations where z does not escape to infinity, do not necessarily lay within the set. The position of an attractor in the complex plane depends on c. The rate of fall through the complex plane is (generally) faster for c closer to the origin.

Bulbs and Periodicity

The main body of the set image is a cardioid. This is usually referred to as the main cardioid. Around its edges are infinitely many "bulbs", which themselves, around their edges, each exhibit infinities of self-similar bulbs. The first order of bulbs are those that are attached to the main cardioid.

There are compound periodicities of orbits associated with the main cardioid and the bulbs. We will call these γ-periods to avoid confusion with other established references to periodicity. Periodicity in an orbit often arises in more than one way. For c values well within the bulbs, there are generally a number of attractors equal to what is generally regarded as the period of the bulb. Similarly for c values well within the main cardioid, there is one attractor, and the main cardioid is generally regarded as having period 1.

However, closer to the edges of these objects, but still within them, the behaviour of the orbit can be more complex. Firstly, strong periodicities emerge in the geometry of the orbit. Secondly, for c values very close to the edges of the main cardioid, for example, the orbit can begin to exhibit multiple attractors. The same thing happens close to the edge of a bulb, as a second-order bulb is approached, and so on.

We are defining γ-periods according to both numbers of attractors and geometric periodicities. The two phenomena are clearly related.

The main cardioid, when c is close to the origin, has an orbital γ-period of 1.

The next biggest bulb attached to the left of the main cardioid, has γ-period 2, and the two bulbs top and bottom, that are next in size, have period γ-3, each when the sea value is well within the associated geometric object.

Moving around the edge of the set to the right from the top and bottom bulbs, the periods of the bulbs follow a series pattern from each bulb to the bulb next smaller in size, round to the valley at the right of the main cardioid.

On this side of the set the periods of the bulbs in this series follow the natural numbers. Between any two bulbs, the bulb next smaller in size in-between them, has a period equal to the sum of the periods of the two larger bulbs.

Moving away from the top and bottom bulbs to the left of the cardioid, the bulbs again follow a series pattern from each bulb to the bulb next smaller in size, but this time, their periods follow the series of odd natural numbers. Again, between any two bulbs, the bulb next smaller in size in-between them, has a period equal to the sum of the periods of the two larger bulbs.

These patterns are only strictly true of c values close to the centre of the objects in question. In the main cardioid and the bulbs, the periodicities of the orbits depends upon the proximity of c to the further bulbs around the edge of the bulb question. In this way, each bulb can be said to have a "basin of influence" that extends around it into other geometric objects.

Escape to Infinity

The set is known to be a subset of numbers on the complex plane where

$$|z| \leqq 2$$

Inside this circle of radius 2 around the origin, we encounter a fractal boundary to the set. It is known that c values on or outside the circumference will always result in z escaping to infinity. There are nonetheless infinite c values within the boundary that are still outside the set, for which z escapes to infinity.

Values of c that do not cause z to escape to infinity are associated with attractors (whose positions are sometimes outside the set), whilst those that escape to infinity from close to the boundary are often associated with obvious repellers. As c values approach the boundary from within the set, attractors can transform into repellers, sometimes in exotic and convoluted stages of transition.

The path of z sometimes orbits around attractors in concentric "rings" shaped "orbit traps", and the dynamics of the path of z as c changes by small amounts sometimes suggest double (or perhaps multiple) attractors within the orbit, as well as the possible coexistence of attractors and repellers inside the orbit.

In other locations the orbit can transition to a set of multiple sub-orbits, each with its own attractor, and as c continues to change these attractors can transition into repellers.

When c is further out from the origin, approaching the boundary from within the main cardioid, but still ≤ 2, the orbit of z towards an attractor can be clearly periodic, or aperiodic, depending on the location of c. Generically, there are "basins of influence" (similar to basins of attraction) within the main cardioid, extending inwards from the bulbs surrounding it, that influence the orbit, and, as it were, compete with each other for influence.

The well-known periodicities of the bulbs, appears to influence the orbit towards their own periodicities, depending on the proximity of c to the bulb, according to two factors. The first factor is the size of the bulb, in which the larger the bulb, the greater its penetration of influence into the interior of the main cardioid. The second factor is the proximity of c to the bulbs in question.

This suggests a picture in which the basin of influence of a bulb not only competes with those of the other bulbs, but also becomes diminished by distance from the bulb.

The result is orbits that descend to an attractor, in various grades of complexity. The simplest descent happens when c is positive on the real axis, and the orbit is in fact on a straight line on the axis, as a diminishing series towards the attractor, which is further along the axis in the positive direction, from c.

When c is negative on the real axis, the attractor is again on the real axis but now closer to the origin than c, and the orbit consists of and oscillation along the real axis that converges to the attractor.

When an imaginary component is introduced into the value of c then the oscillation becomes "spread out" along the imaginary dimension, resulting in the path of z spiralling in to the attractor, which also then has an imaginary component to its position.

The geometric nature of the spiral depends on the position of c and the influence of the bulbs. The valley on the right of the main cardioid is a zone with γ-period 1, "causing" single armed spirals, whilst the largest bulb to the left, "causes" two-armed spirals. Further away from the real axis but still well within the main cardioid the influences from the other bulbs create orbits in the form of multi-armed stars, star polygons, and multi-armed spirals (stress-rotated armed stars) that emerge from rotated star polygons.

The patterns of relations between the periodicity of the bulbs include the progression of the natural numbers, and the Fibonacci series. Other significant numbers occur in the relations between features of the set, such as the transcendentals π and e.

The most interesting behaviour in the orbit of z happens close to the boundary of the set and just outside the boundary. In one scenario, the number of arms of the stars and spirals in the orbit of z increases dramatically as c approaches the boundary, as it crosses lines of influence of the bulbs, and is influenced by the smaller bulbs with higher periods.

The attractor at the centre of the orbit for that c value changes form as c moves across the boundary. Attractors can cause rings, polygons, foils, and so on, of increasing size. As this happens multiple new repellers can appear in place of the vertices of the arms of the spiral or star, from which z values then escape to infinity as c moves outside the set.

In another scenario where c crosses from the main cardioid into a continuation of the set in one of the bulbs, the orbit of z forms an n-armed star, until the central attractor causes a ring or an n-polygon whose vertices are the z values cycling around it. Further into the bulb, z orbits in a way close to being a closed irregular point orbit. As the boundary of the *bulb* is approached by c, and then c moves outside the set, the attractor changes to a repeller.

Highly exotic orbits of both escaping and non-escaping iterations of z, involving multi pointed or armed stars, spirals, and dense, closed orbital forms of various shapes, can be observed in certain locations at the boundary of the set. Such exotic behaviour of the orbits of z become generally more pronounced closer to the boundary of the set, whilst orbits well within the main cardioid are relatively straightforward and become more so, the closer c is to the origin.

Discussion

The pre-requisite for the Mandelbrot set object (as it is currently understood), is the complex plane. The only real numbers that appear in the set are those on the real axis, however, it is an easy step in principle to convert complex and imaginary values to real number positions on a real plane. This is what computer images of the set do. The actual set, however, in order to arise through the iteration equation, *requires* complex numbers.

We can ask the question "are imaginary numbers genuinely objective"? The fact that they are called "imaginary" is misleading. Gauss preferred to call them "lateral" numbers. We cannot find imaginary quantities in nature *explicitly*, but they are nonetheless widely implicit in natural phenomena.

Even if we wanted to argue that imaginary numbers in themselves are not genuinely objective, then their usefulness, and in many cases essentialness, in mathematics and science, and engineering, and consequently in the creation of many technologies, is sufficient demonstration in itself that there is *something* about them that is genuinely objective. Specifically, their implicit existence in mathematical structures is genuinely objective.

So we can that in natural phenomena they are genuinely objective in an *implicit* way. That is to say, they *implicitly exist* in the *working* of natural phenomena as *we scientifically understand the phenomena*, but all the natural quantities that we *actually encounter* in a physical way such that we can quantifiably measure them, are quantities expressed by real numbers.

Imaginary numbers are a prime example of how there is not a clear dividing line and separation between the *processes of our understanding* of mathematical structures a natural phenomena, and the mathematical structures and natural phenomena that we are understanding. There is not a definite separation between our intellect and intelligence and ability to reason, especially mathematically, and the objects to which we are applying this capacity. This is the case whether they are objects in the mind, or objective structures of material phenomena.

Fractal Structure

In mathematics the complex plane is usually considered as a 2-dimensional continuum. In fractal geometry a fractal object is often said to have an approximated fractal dimension D:

$$D = \frac{\log n}{\log s}$$

which may be defined in various ways for different types of fractal. One example is where n is the number of objects at a given resolution, at a magnification s, either for the whole object or for some limited part of it.

It is questionable whether the idea of fractal dimension has other than simplistic applicability to the wider phenomena of nature in general. We use fractal generating processes now, for creating "realistic looking" landscapes and natural features, in computer imagery for virtual environments. Such techniques are typically overlaid and complex, and in general, when it comes to the "real world", we can expect the variety and complexity of nature herself to be characterised by principles far more complex than what can be described through the concept of single fractal dimension.

Within the terms of Object Theory we can look very easily at the concept of fractal space. When we conceive a space object as a continuum, such as, for example, the complex plane, or even just everyday Euclidean space, we have already instantiated an object. Let us just called this object "the space".

Now we come to put other objects "in" that space, or to say that they are subsets of that space. Any fractal object will do for illustration. There are plenty of examples, but let us just take the very simple example of a "cross fractal".

We start with a square in the Euclidean plane, and draw a centralised cross (first iteration):

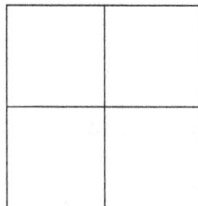

Then we repeat for each square made (second iteration):

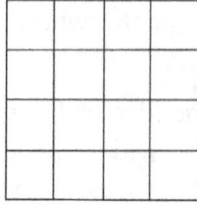

We the repeat again, for each now square made, and so on, with infinite recursion. According the equation above for fractal dimension this theoretical fractal has a dimension of 2, the same as that of the Euclidean space, which also happens to be about the same as that of the Mandelbrot set, which is incomparably more exotic and interesting. It is an example of how fractal dimension is not an index of exoticness or interest!

We can represent this infinite, fractal "graph paper" object as an IIP:

$$\circlearrowleft (\infty) \{ \boxplus \}$$

where $\boxplus$ is the process for filling all squares with crosses. If we start with unit square then over finite iterations as

$$\circlearrowleft (n) \{ \boxplus \}$$

the area of each square is simply

$$\frac{1}{4^n}$$

In contrast the contemporary conventional understanding is to say that for the *infinite* iteration

$$4^n \longrightarrow \infty$$

and so

$$\frac{1}{4^n} \longrightarrow 0$$

Page 162

and therefore the area the area of each square is then *infinitesimal*. In the way we do mathematics we have conveniently developed the concept (an object) called "nonstandard" or "hyperreal" numbers or quantities that are closer to zero than any "standard" real number, but are not zero.

What we have done here with the cross fractal is essentially to create a coordinate system for the square. It is of course a theoretical construct, just as is the concepts of the Euclidean "point', "line", and so on. A coordinate system or indeed the Euclidean plane itself is conceived as being filled with infinite "points", just as the axes of a Euclidean coordinate system are conceived as being filled with the "infinity of the real numbers".

But if we are to be consistent then by definition no real number is as small as a "nonstandard" hyperreal infinitesimal number. And if both real numbers and hyperreal infinitesimals exist on the grid, then the reals must be separated from the continuum somewhere, by the infinitesimals. That in turn means that we cannot use (standard) real numbers to express *any* position on the continuum.

Given that positions are relative, this means that with the presence of infinitesimals, even rational numbers and natural numbers cannot properly be expressing positions on the continuum of the plane.

Of course we *know* we *cannot use* real numbers to express the areas of the squares, because no real number can be both literally infinitely small, *and* a finite number. And of course, any finite number, however small, will tend to infinity as its multiple tends to infinity.

There are all kinds of problems that one can construct, just as there are all kinds of answers and conventions that one can construct to overcome the problems. There are constructs, and the objects called infinitesimals and hyperreals is one of them, that have been invented specifically to handle these problems. And as tools, they work. Mostly.

We need to address here, the contentious issue of invention versus discovery. It is true to say that we invented such objects, but it is also true to say that we discovered them. From the point of view of Object Theory both of the same thing. They are discovered in the workings of our mind, and our comprehension. Like, for example, the so-called "imaginary" numbers (that Gauss called "lateral"), they are indeed "made up" through the use of our mind and intellect, but that does not mean that we have not *discovered* them in our world. We have still discovered them, because our mind, arising through brain function, is *part of* our world. And furthermore, our world itself, whatever we know of it, or experience of it, or work out about it, using mathematics, is only ever what arises as a construct of our brain function.

So contemporary mathematics is full of constructs. What "traditional" mathematical "rigour" doesn't do, in assessing these constructs, is acknowledge or take into account the construct that lays behind all such assessments, which is the *construct of brain function* that the intelligence we are using *is*, and through which we experience the phenomena of the world.

Such constructs, if they are "rigorous", are constrained by a network of relations to what has already been established in the same way. In this way, in such constructs we are not necessarily looking at something that is genuinely objective, but rather, something that is consistently objectively valid, in relation to this network, at the foundation of which, is always certain "unprovable" axioms.

Ultimately, "unprovability" or uncertainty, arises in connection with the tacit belief or assumption that we are applying our intelligence to something separate from the intelligence we are using. So if that belief or assumption is false, even when it is only tacitly present, then "unprovability" or uncertainty must remain because it is already inherent in the fact that the nature of the mind or intelligence itself, that we are using, has not been grasped).

We are saying now that we can construct a fractal Euclidean coordinate system as the IIP:

$$\circlearrowleft (\infty) \{ \boxplus \}$$

We are *not* saying we are constructing a "grid of the real numbers" under infinite iteration, but rather, simply a Euclidean grid by binary divisions.

We can go further. The distance D between adjacent lines on each iteration is

$$D = 1/2^n$$

on the nth iteration, and it is perfectly possible to *number* the lines, by adding a line numbering step into the process. So, in addition to the process $\boxplus$ of filling all squares with crosses, we also have a process that on each iteration simply deletes all the line numbers from the previous iteration, and renumbers all the lines outwards from the origin (origin = 0), with the natural numbers.

So the IIP now we could write as, for example:

$$\circlearrowleft (\infty) \left\{ \begin{matrix} N \\ \updownarrow \\ \boxplus \end{matrix} \right\}$$

We then have that the distance M from the origin of any line with line number m, will be

$$M = m/2^n$$

M is now the actual *real number* ordinate value for that line.

Here, note that because we are confined to the unit square, m is in any case constrained by

$$m \leq 2^n$$

Given that M is a real number ordinate, what happens as the iteration "goes to infinity"? We must conventionally conclude that as

$$n \longrightarrow \infty$$

then also

$$M \longrightarrow 0$$

but also, if we look carefully, either that:

Page 165

M values (which are always real numbers) $\longrightarrow$ all real numbers between 0 and 1,

or

the size of the infinity of reals between 0 and 1 is not the same as the size of the infinity given by the expression

$$m/2^n$$

when $n \longrightarrow \infty$.

These are the kinds of notions that arise when we try to conceive the object "infinity" through the same concept structures we use when conceiving multiplicities of distinct objects, as objects called "sets".

In fact, there is no necessary compulsion to conceive the meaning of

$$\circlearrowleft (\infty) \left\{ \begin{matrix} N \\ \updownarrow \\ \boxplus \end{matrix} \right\}$$

in this way at all. Nor indeed is it necessary to conceive the meaning of the more conventional expression

$$m/2^n \longrightarrow 0 \text{ as } n \longrightarrow \infty$$

in any way that implies we are expressing a situation in which infinity is *actually involved*. These are merely *informal* expressions that *exclude* the actual infinite case, and merely show that when n becomes extremely large, then

$$m/2^n$$

becomes so small that it can be considered negligible.

There *is no case* in which the term

$$m/2^n$$

has any meaning when n is infinite, because as a piece of algebra *it represents a number structure*, that applies to number of the kind that such structures apply to, and are meaningful in relation to. If we substitute for n, "infinity", then we have not substituted a number of this kind. "Infinity" is not a number of the kind that such structures of operation apply to.

In the grid, the distance of the nth line from the origin is always:

$$N = n/2^n$$

For n iterations greater than one, there will always be lines between this line, and the edge of the square.

Now a number of different things are *independently* or *separately* implied just from the equations, as $n \longrightarrow \infty$. These are:

$$D \longrightarrow 0$$
$$M \longrightarrow 0$$
$$N \longrightarrow 0$$

If we again think this says something about an actual result of infinite iterations, then again we will be misled. This is where we would have to introduce the concept of infinitesimals if we are to avoid a situation in which our construct implies that the diameter of the square as $n \longrightarrow \infty$ becomes undefined or zero, and that the nth line has no lines between it and the edge of the square.

Alternatively we can abandon the idea that we are ever even looking at or representing the case of infinite iterations in this way.

We can illustrate this. In actuality, such as where we actually construct this on a computer, however much we zoom in on this fractal, we are always still in the finite domain, and so in practice this problem does not arise. And yet still, we may in principle zoom in infinitely. Anyone familiar with computer generations of fractals will be familiar with this principle.

Our recent discovery and knowledge of fractals, in particular our experience with them through computer processing, has literally presented fractal objects to us, as *processes*.

Page 167

In the "real world" situation the suggestion that the conceptual object "as $n \longrightarrow \infty$" represents some situation that is somehow different to an extremely large but nonetheless still finite "n", has no meaning. It is an intellectual artefact. The IIP

$$\circlearrowleft (\infty) \left\{ \begin{array}{c} N \\ \updownarrow \\ \boxplus \end{array} \right\}$$

is not a process that converges in the way that, say, convergent infinite series or integrals do. It is always a *process*, and as an object remains a process, always, even when we examine a halted condition of it.

Now the same is true of the natural numbers, and the real numbers. And given natural number is a halted condition of the IIP that produces natural numbers. Any real number is an IIP that is the process for producing that number, as a structure of natural numbers.

In mainstream mathematics, however, the current predisposition is to consider all mathematical objects - unless they are explicitly described as processes such as infinite series, or iterations - as objects that are devoid of process. In truth, every single object that we encounter in mathematics is a process or a part thereof.

It is common to conceive the Mandelbrot set, for example, as an "infinite set", which means we imagine it to be a fixed and complete object of a kind that is not a process. It is as if we think we see a "brick" of infinite interest because it is composed of infinite parts, with infinitely interesting relations between themselves, but is nonetheless a brick. We use a *process* to discover it, and then imagine that what we have discovered is not a process, because we conceive it in the context of the constructed tower of conceptions in contemporary mathematics, built on the concept of the "set", and understand it in that way.

Set theory analysis contains certain limitations due to the way the conceptualisation process is happening in the theory. It is not widely recognised that one of these limitations arises from the fact that set theory necessarily deals with *distinct* objects. This precludes the ability to be able to understand and describe situations where distinction is only an illusion. Such situations do arise, and they have indeed arisen in set theory itself, where they have resulted in the development of the formalisations necessary to avoid inconsistency or paradox.

Structure in a Fractal Continuum

Let us define two IIPs:

$$S_1 \equiv [\circlearrowright (\infty) \{Q_1\}]$$
$$S_2 \equiv [\circlearrowright (\infty) \{Q_2\}] \quad (1)$$

A simple structure between them is then:

$$S_1 \longleftrightarrow S_2$$

Let us define the relation object between them as "S_1 is identical to S_2 when scaled by a scaling factor M". Let us call this relation F_M. So:

$$S_1 \longleftrightarrow F_M \longleftrightarrow S_2$$

We can do the same thing again, adding in a third IIP called S_3 :

$$S_1 \longleftrightarrow F_M \longleftrightarrow S_2 \longleftrightarrow F_M \longleftrightarrow S_3$$

We can do this indefinitely, for any number of IIPs between S_m and S_n , and write it simply as:

$$S_m \leftarrow \frac{F_M}{} \rightarrow S_n$$

As it stands, this refers to a specific scaling factor, M. Now let us allow other scaling factors that are not F_M but arranged so that they are periodically $M^\wedge n$. For example, if M is 2, then the relation is:

$$\text{periodic scaling factor} \quad 2 \longrightarrow \quad 2 \longrightarrow \quad 2 \longrightarrow \quad 2$$
$$\text{magnification} \quad 2X \longrightarrow \quad 4X \longrightarrow \quad 8X \longrightarrow \quad 16X$$

Now let us also say that the scaling factor is a "sliding scale" or a continuum. So we can write this very simply as:

$$S_m \leftarrow \overbrace{Z\{M\}}^{\circlearrowleft} \rightarrow S_n$$

where

$$\overbrace{Z\{M\}}^{\circlearrowleft}$$

means any scaling factor or "zoom" that is periodically $M^\wedge n$, where n is an iteration of

$$\overbrace{Z\{M\}}^{\circlearrowleft}$$

We can further define

$$S_m \leftrightarrows S_n$$

as a two-way *morphism* between S_m and S_n that takes place under the scaling

$$\overbrace{Z\{M\}}^{\circlearrowleft}$$

and under which S_m changes to S_n over some kind of continuum, such that S_m and S_n are identical.

It will be pretty obvious by now that what we are describing is a perfectly self-similar fractal of some kind.

What if we wanted to describe a fractal that is not perfectly self-similar? Then we could simply define the morphism

$$S_m \leftrightarrows S_n$$

as one in which S_m and S_n are not necessarily identical, but are nonetheless self-similar.

Now what happens if, by the terms of Object Theory the relation between the original two objects turns out to be that Q_1 and Q_2 are not really two distinct objects?

$$Q_1 \leftarrow \frac{\text{false}}{} \rightarrow Q_2$$

Then by the terms of Object Theory S_m and S_n are only distinct objects in that they are distinct *instantiations* of *one of the same object*. That object being the original object that became instantiated as these objects S_m and S_n.

And so now, the structure

$$S_m \leftarrow \overbrace{Z\{M\}}^{\circlefthalf} \rightarrow S_n$$

needs to be written more faithfully, in terms of the original one object, as, for example:

$$\{Q\}_A \leftarrow \overbrace{Z\{M\}}^{\circlefthalf} \rightarrow \{Q\}_B \quad (2)$$

where

$$\{Q\}_A \text{ and } \{Q\}_B$$

are each some kind of distinct *instantiation* of the *one object Q*.

Now of course (2) above, which shows the morphism between *instantiations*, is also a structure, and so can additionally be represented in the usual way of representing a structure as:

$$\{Q\}_A \longleftrightarrow \overbrace{Z\{M\}}^{\circlefthalf} \longleftrightarrow \{Q\}_B$$

Page 171

meaning that the object

$$\overbrace{Z\{M\}}^{\circlefthalf}$$

here, is the relation between the two distinct *instances* of one and the same object Q. The instances are created as distinct instantiations in the first place, precisely by putting the scaling continuum as the *relation* or the object, *between* these two instantiations. In other words

$$\{Q\}_A \text{ and } \{Q\}_B$$

are distinct instantiations *because of* the scaling continuum.

The original object Q, which is an IIP, may or may not be a *quantity continuum*. It may be some other structure. Either way it is what the object we are calling a "fractal", *is*.

Let us now consider the case where Q is indeed a quantity continuum, such as, for example, the complex plane. It is an IIP to which we apply the scaling object, and under scaling, it undergoes a morphism exhibiting self-similarity. The complex plane as a continuum is of course perfectly self-similar at every scale, but if we express it with numbers, then it will exhibit periodic self similarity with the periodicities of the numbers used to express it.

Mainstream contemporary mathematics conceives the fractal object as a set. "Set" is an object in its own right. However, "set" is an object that cannot be satisfactorily used to describe a continuum, such as the complex plane, for example, without entailing other constructs such as non-standard real numbers, in the way that we have discussed.

The underlying reason for this (in terms of Object Theory) is that the standard concept of a set necessarily entails the concept of it consisting of *distinct objects*. In contrast, we do not necessarily need to conceive a continuum quantity as also consisting of distinct objects, if we do not take the continuum quantity and the numbers used to express it, as one and the same object.

Page 172

So we are saying that by *relating* two instantiations of Q through the scaling object, Q is fractal if through the scaling relation it exhibits self similarity. If Q is a continuum such as the complex plane, for example, then it will exhibit self similarity by virtue of the self similarity inherent in the scaling object.

There is a relation between the scaling object and the quantity continuum:

$$\overline{Z\{M\}} \overset{\circlearrowleft}{\longleftrightarrow} \circlearrowleft (\infty) \{Q\}$$

What is the generic form for *any* such scaling object, rather than this one with a specific scaling factor M. We can write a scaling continuum IIP as:

$$\circlearrowleft (\infty) \{Z\}$$

This is a "sliding scale" scaling *continuum*. Precisely because it is a *scaling* continuum, it is periodic in the same way that natural numbers are periodic, and this periodicity is an iteration process. In other words, it is a "sliding scale" magnification or zoom factor, available at any zoom factor, but as we zoom, periodically the zoom factor will be $2, 3, 4$ and so on.

So now in applying it to Q, we are talking about a relation and a structure that is:

$$\circlearrowleft (\infty) \{Z\} \longleftrightarrow \circlearrowleft (\infty) \{Q\}$$

If we want to describe any particular scaling factor on this scaling continuum, we have to apply a number to it. Let us be able to express any scaling factor as a real number, so we will have the scaling structure:

$$\circlearrowleft (R) \{\circlearrowleft (\infty) \{Z\}\}$$

The scaling continuum is now calibrated with the real numbers. Now we can apply it to the continuum object Q:

$$\circlearrowleft (R) \{\circlearrowleft (\infty) \{Z\}\} \longleftrightarrow \circlearrowleft (\infty) \{Q\}$$

If we want to take a measurement on Q, and relate it to the measurement on the scaling continuum, then we need to apply numbers to Q. Then we can have the relation:

$$[\circlearrowleft (R) \{\circlearrowleft (\infty) \{Z\}\}] \longleftrightarrow [\circlearrowleft (R) \{\circlearrowleft (\infty) \{Q\}\}]$$

One such relation is the fractal dimension. For example if we re-write this for simplicity as

$$[A] \longleftrightarrow [B]$$

then the relation that is the fractal dimension might be given as

$$\frac{\log [B]}{\log [A]}$$

provided that Q is the kind of quantity to which we can apply a real number, and then take the logarithm.

However, the concept of "fractal dimension" is limited. In general, as we scale a nonlinear object by "zooming" there are a number of separate features that can change with the scaling factor. The degree of complication is only one feature, and self-similarity that is also a matter of degree is a separate feature. There is also the relative relations of component objects, within the geometric space, which may or may not be self-similar to some degree, great or small, depending on the zoom. They may be self similarity at different scales, that is rotated, for example.

Fractals in general are not necessarily "perfectly fractal" meaning, they are not perfectly self similar at every scale. Sometimes this kind of self similarity is observable. In other cases, the fractal object undergoes morphisms with scale.

One of the features particularly exploited by computer imagery of fractals such as the Mandelbrot set is that the *rate of escape to infinity* close to the boundary of the set, can also be fractal. Even the *rate of escape* can be considered in much the same way, as a structure of continua.

Julia Sets and Numbers

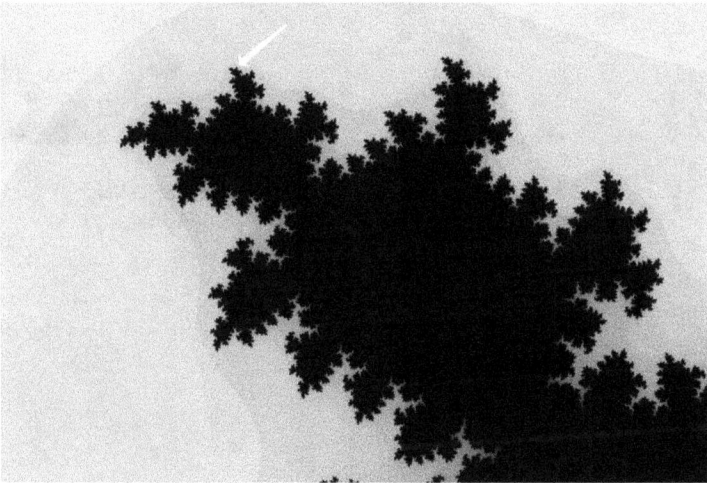

PART OF THE JULIA SET FOR $Z_0 = 0.636$, C = 0, ZOOM 21

**THE SAME JULIA SET, ZOOM 31, POSITION SHOWN
BY THE ARROW IN THE PREVIOUS IMAGE:**

Now let's apply this approach to the Julia set for $c = 0$. The iteration function is

$$z \longrightarrow z^2 + c$$

or in our adopted notation:

$$\circlearrowleft (\infty) (z \longrightarrow z^2 + c)$$

A "filled" Julia set is any set of initial values z_0 for which z in the infinite iteration remains bounded.

When c is at the origin then it is well known that the set comprises a circle around the origin. This is the simplest Julia set, and would not usually be described as fractal.

The circumference of the circle is a continuum quantity. In our description

$$[\circlearrowleft (R) \{\circlearrowleft (\infty) \{Z\}\}] \longleftrightarrow [\circlearrowleft (R) \{\circlearrowleft (\infty) \{Q\}\}]$$

Q could now be the circumference of the circle. It is self-similar, but gradually changes to become straighter (relative to the unit size) as the scaling factor increases.

A *connected* "filled" Julia set is not a direct consequence of the iteration function. It is a consequence of the *selection choice* applied to the orbit of z that is the *result* of the iteration function. The selection process is that the set only contains those initial z values for which the iteration *does not escape to infinity*. We will talk about disconnected Julia sets later.

We could symbolise a structure for the actual IIP producing the set as:

$$\circlearrowleft (z_0) \{C\} \overbrace{\{\circlearrowleft (\infty) \{z \mapsto z^2\}\}}^{\uparrow|\checkmark}$$

Here, above the brace is the combined symbol "up arrow, OR, tick", meaning that if the infinite iteration below the brace escapes to infinity then the z_0 value in the iteration on the far left is rejected, but if the iteration below the brace remains bounded, then the tick applies, and the z_0 value in the iteration on the far left is included in the set. On the left is the IIP producing a complex number z_0 applied to the complex plane continuum C.

Now we come to a core issue: There does not exist any IIP for creating "all" numbers on the complex plane, just as there is no IIP for creating "all" real numbers. So what is this IIP $\circlearrowleft (z_0) \{C\}$ that we are using?

***

Let us take a detour to fully look into this now. First, let us be clear about the infinites of distinct objects we call numbers. We may speak of "all" the *natural* numbers as an IIP. As long as we realise we are talking about a *process*, and not an object that is something other than a process, then we are not talking about something potentially problematical. If, alternatively, we conceive of an object we call "all" the natural numbers" as something that is an object or a structure that is not a process, then we have created a chimerical concept.

The natural numbers are an infinity of distinct objects, that are created through the natural number creating IIP. That means, "all the natural numbers" is an object that is an infinite, unhalting or an ending *process*. We have already talked about this at length. No natural number can be considered as an object with no relation to the IIP. All natural numbers are part of the IIP. It may seem initially counterintuitive to consider a single natural number as inherently a process, or part of a process, but that is what it is.

Mundanely, it appears as a "static" object, rather than being object that has something to do with processes. The same is true of a computer program appearing as a page of static symbols on a piece of paper or screen. But that is only a mundane appearance. The importance and meaning is in the process. In the case of the computer program, it is in the *processing* of the algorithm. In the case of the number, it is in the processing of that number's relations to other numbers in a structure.

There *is* an IIP for creating the natural numbers, and we know what it is. We know all about it. It exists as something that we can define. We can show what it is, and how it works, for whichever number base we choose. However, we cannot take the same approach with the real numbers, or the numbers on the complex plane.

There is a very good reason for this. All real numbers are constructed from components that are natural numbers. The natural numbers are an infinity of distinct objects. Let us consider for example real numbers between the natural numbers 1 and 2.

- In decimal, any such numbers will begin with a "1" followed by the decimal point.

- In the case of an irrational number there will infinitely many digits after the decimal point - the IIP does not halt.

- Each real number, rational or irrational, is generated from an IIP. In the case of a rational, the IIP continues to produce zeros after the last non-zero digit, or an infinitely recurring set of numbers.

- In order for two such numbers to be distinct, there must be at least one place position after the decimal point where different digits occur for each number.

- Therefore, the IIPs for distinct irrational numbers must themselves be distinct.

- There does not exist an IIP for producing consecutive real numbers for the following reason:

- Let two distinct real numbers be generated by the IIPs I_1 and I_2, where the real number generated by I_2 is greater than that generated by I_1.

- If I_1 and I_2 are both rational then they are not consecutive. So let I_1 be irrational.

- I_1 is therefore an *infinite iteration* process and therefore does not halt. Therefore, there can be no iteration process I_3 in which I_1 is an iteration whose result is used as the initial value for the iteration process I_2.

- If we let I_2 be irrational, or I_1 and I_2 both be irrational, then the easy to find corresponding argument applies.

So there is no IIP that we can construct that will produce consecutive real numbers, in the way that we can indeed construct an IIP for producing consecutive natural numbers. And this is consistent with the fact that there is no such thing as "consecutive" real numbers with no real numbers "in between". In recognition of this problem Gauss constructed the object we call hyperreal numbers, to get around it.

Every real number is itself the output of an IIP. In the way that we discussed above already, even a rational number, as a terminating decimal, can have infinite zeros appended to its last digit, which means it, also, can be considered as an *infinite* iteration process. Alternatively it can take the representation ending in infinite 9s.

However, the *basis* of the real numbers, the objects from which they are constructed, are the *natural numbers*. Real numbers are expressed using natural numbers. And the natural numbers are the labels that we affix to the infinity of distinct objects - an object that exists in our mind, but that we are unable to demonstrate is genuinely objective.

So when we are creating real numbers in general, and correspondingly if we are creating numbers on the complex plane, then we are engaging in applying the natural numbers - which are "labels" for an infinity of distinct objects - to an iterative process in which there is iterative *scaling* of the quantities those natural number stand for, that corresponds to the iterative positions before and after the point.

The natural numbers in themselves always only stand for quantities of distinct objects, or what is sometimes called discrete quantities. Using them to express continuum quantities requires scaling the quantities that they stand for. However, ultimately, to be able to properly express all continuum quantities, using constructs of natural numbers called real numbers, it is necessary sometimes for this scaling *process* to be *infinite* and unhalting (as in the case of an irrational).

Without recognising that all real numbers are the *creations of processes*, and could be understood in terms of networks of processes, rather than structures of static concepts, then it is still indeed possible to build and evolve entire structures and networks of conceptual objects and representations, through which we can go ahead and understand the nature of numbers. At present, this situation exists in the contemporary understanding of numbers and their relations through set theory, and other concepts such as the concept of the field.

Nevertheless, this entire understanding is a structure of mind, operating in a particular way. It is not difficult for entire networks of mind to come to work in the same way, through the principle of learning, and such learning is in principle a learning *how to think*, and *how to reason*, based on principles and operations of mind, and mental processes that in themselves, are not necessarily consciously known or seen. The principle by which this becomes established in *networks* of minds, is in fact the principle of neuroplasticity working in networks of brains.

Brain function itself, and the mind and reasoning capacity that we are using in mathematics, is a *process*. Up until now, mathematics and science has proceeded on the basis of understanding something separate from the mind being used to do the understanding. It is not possible to get beyond this phase, by coming to understand something new, some new theory or fact for which we have "proof", where that thing is still something separate from, or seen as separate from, the mind that is doing the understanding.

So what we are doing here is not providing some "proof" in the traditional way, of some thesis about numbers and their relations being processes, but rather, we are providing pointers for what is literally a new psychological approach.

Up until relatively recently in its history mathematics has been widely regarded as an ivory tower repository of facts and proofs and various means of understanding that supposedly transcend the subjectivity of human psychology. In truth, this is a relative situation, a situation that has been changing, and will continue to change, as the world changes, as computing power increases, as artificial intelligence becomes more powerful, as physics makes new discoveries, as knowledge of the human brain increases, and as the psychology of human beings and their methods of reasoning changes and comes under the spotlight of neuroscience. As all this happens, mathematics itself cannot maintain its previous position. And it is already showing signs of change, despite where there is attachment to the past.

So in the past there have been competing mathematical arguments and theorems for dealing with the concept of infinity. And always, traditionally, these have been presented as structures assumed to be objects that we can reason about, or construct arguments for or against, that are treated as separate from the processes of mind that are being employed to carry out this activity. The direction of attention has always been outwards, from the mind, onto the object, as a kind of attempt to externalise the workings of the mind as structures of objects.

Where we are heading, it will be necessary for this direction of attention to turn inwards, as well as outwards. The assumption that the inner workings of the mind can be discovered by projecting its workings outwards into object oriented structures, is just an assumption. And it is one that is mistaken.

As it turns out, the concept of "all" the numbers on the complex plane, *even if conceived as a process*, is chimerical. Of course it is possible to *conceive* of the *"set of all values on the complex plane"*, and this would indeed be done within the bounds of set theory. It certainly makes superficial sense if we don't look too deeply into it.

The concept of "the set of all complex numbers" is a very commonly encountered concept indeed, in contemporary mathematics. But in terms of Object Theory that structure, which we could write, if we wish, as:

$$["set"] \longleftrightarrow ["all" \longleftrightarrow ["values\ on\ the\ complex\ plane"]]$$

is a *intellectual construct* that has its basis in a psychology of reasoning, that connects the concept of "all" that is perfectly valid when applied to certain finite quantities, into this structure *illegitimately*, because the structure contains an implicit infinity, or reference to an infinity, in the object "all", that it does not declare. How the transition from the "all" concept applied to a finite plurality, to its application to an infinite one, is not explained.

In order to analyse the concept of "the set of all points on the complex plane" properly, we would have to get into the structure of that object "all". Because that is where the sleight occurs.

The inductive leap inherent in this construction does not, however, prevent tools of analysis and understanding from being constructed out of it. It does not render set theoretical analysis based on the concept any less useful. We are simply saying that the object "the set of all points" is a feature of mentation that has no genuinely objective counterpart. And the fact that one can build structures of concepts and reasoning, that appear to "prove" the existence of the chimera, does not mean that such a chimera exists genuinely objectively.

What we have been saying is that the complex plane itself, however, is a different matter. The complex plane itself, too, as an object that we conceive, is a construct. What we understand it as, is a mental construct created out of human intelligence. In these times we address it through the object we call a set, but it is an object that we can conceive as a *continuum quantity* and treat its infinity in a different way. Not because we are claiming anything special about a *continuum* that isn't ordinarily understood about it, but just because as an object it is not the same thing as the object we call a number, or indeed, the object that we called the "infinite set of all numbers on the complex plane".

We can write it as an IIP:

$$\circlearrowleft (\infty) \{C\}$$

where C stands simply for "complex plane continuum", which is also a theoretical construct, but a very useful one, in retraining how we are thinking, because it does not inherently contain in itself, any reference to *numbers*. We can also consider numbers themselves, as infinite iteration processes. We can ultimately suspect that there will be number relations between structures, in the sense that we can apply numbers as a tool, to understanding structures and their relations.

We said that there is no IIP that we can use for creating all the real numbers or correspondingly all the numbers on the complex plane. We have just illustrated how we should not be surprised by this, given that the concept of "all" the numbers, is an intellectual construct to begin with, that we can show is chimerical. It is not the fact that it is a chimera that is the problem, the limitation is in that its chimerical nature has not been detected and understood. Its use allows the building of structures of reasoning arising from psychological processes that have not been understood and recognised.

The IIP for producing the natural numbers is of one fixed kind, for each number base. Even children know it in the everyday way as a counting system. However, there are many different IIPs that may produce a given real number. When we have a set of equations, for example, these IIPs are inherent in the mathematical operations and relations that the equations and their relations consist of. What we are ultimately dealing with, is structures of relations of processes.

***

So in the IIP structure that we gave for producing the Julia set, what do we mean by

$$\circlearrowleft (z_0) \{C\} \text{?}$$

What kind of an IIP could this be if it isn't an IIP for creating "all" the numbers on the complex plane. There are a number of ways we could interpret it, one being that we can take this to stand for an algorithm that puts a z_0 value into the iteration process on the right, and is also informed by the accumulation of z_0 values that enter the set. Another z_0 value is then created, randomly if we wish, and only rejected if it already exists in the set.

In terms of set theory, on each iteration of this IIP the value of z_0 is taken from outside the set of all sets of z values that the iteration of z has so far produced.

So the IIP as shown:

$$\circlearrowright (z_0)\, \{C\}\, \overbrace{\{\circlearrowright (\infty)\, \{z \mapsto z^2\}\}}^{\uparrow | \checkmark}$$

is a structure for the creation of the Julia set for $c = 0$, with the iteration process on the left taking any one of many different forms. The iteration is essentially just that of "picking" a new number from the complex plane, on each iteration. *Whatever form it takes, it won't affect the outcome.*

The general structure is:

$$\circlearrowright (\infty)\, \{C\}\, \overbrace{\{\circlearrowright (\infty)\, \{z \mapsto z^2 + c\}\}}^{\uparrow | \checkmark}$$

where c is a constant.

The actual iteration function itself, does not produce the connected Julia set without the intervention of the $\uparrow | \checkmark$ principal that separates the z values that escape to infinity from those that do not.

Nevertheless, the most salient feature of the Julia set, is the *iteration of form* that we call "self similarity". If we represent the structure simply as J, then we have that *self similarity* in J, at different scales, arises from the relation of J to infinity, as the structure:

$$J \longleftrightarrow \circlearrowright (\infty)$$

and there is therefore an object O in:

$$J \longleftrightarrow 0 \longleftrightarrow \circlearrowleft (\infty)$$

of interest, that we should be looking to understand.

The fractal features turn out to be much more salient for c values on the boundary, and just outside the boundary of the Mandelbrot set, where Julia sets become much more interesting, and change from being connected to disconnected. All of this interesting structure is *clearly a manifestation of iteration processes*. However, it is *not* directly the iteration

$$z \longrightarrow z^2 + c$$

that is the cause.

This principle of iteration is illustrated in the images at the beginning of this section, where we can see how the edges of the characteristic forms of the components of the object consist literally of reiterated forms of the same kind, but in a different scale. This process of (re)iteration is infinite, and in this particular example the iterations change the form very little over many iterations.

15: The Natural Continuum

L et us review some basic principles we have been using. The fundamental (and abstract) IIP object, before being furnished with any other properties, we have been writing as:

$$\circlearrowleft (\infty)$$

If we now attach this to a continuum Q, as

$$\circlearrowleft (\infty) \{Q\}$$

then what we have written is an infinite iteration process attached to a quantity object. More usually, if we just write Q, then that stands for a quantity as a number, which still *implies* an iteration process for that number or measurement, but one of discrete numbers or distinct number objects.

The expression we are using does not yet refer to a number. Also it does not represent an "infinite quantity", but rather, one could say, an infinite *supply*. The supply of the quantity, whatever size the supply, has to happen through a *process*. That process is our infinite iteration or a part thereof. More usually, we would use *numbers* or letters representing numbers, to represent this supply.

An infinite continuum of some quantity such as "distance" is something that provides an infinite supply of quantity, of whatever the "stuff" is that it is a quantity *of*. We are representing the principle of supply detached from any specification of any particular "stuff", such as "length", "time", or "energy", and so on. The supply principle for a continuum quantity is the principle we have encapsulated as the IIP.

Another example: Whilst in a graph we might ordinarily say that the y axis is infinite, we are saying here that there is an infinite iteration process, an IIP, that supplies the values that the y axis represents. So we might write that as

$$\circlearrowleft (\infty) \{y\}$$

We can then do the same thing for the infinite x axis:

$$\circlearrowleft (\infty) \{x\}$$

Normally, would say that the axes define a *space* that is a *set* of *points*. We could then use set theory as a tool to explore features of objects that are a subset of this space.

In contrast, we are saying that the two IIPs have a *relation*:

$$\circlearrowleft (\infty) \{y\} \longleftrightarrow \circlearrowleft (\infty) \{x\}$$

They are two distinct instances of the object called distance or space, and the relation in the first instance is that they are *orthogonal*. In the case of Euclidean geometry this means at right angles, but because we are working outside that conception, orthogonal means essentially that an operation or structure involving one of these IIPs does not affect the other. The relation is one of *disconnection*.

A function $y = f(x)$ is a process object that here becomes a relation between the two IIPs, as:

$$\circlearrowleft (\infty) \{y\} \longleftrightarrow [\text{function}] \longleftrightarrow \circlearrowleft (\infty) \{x\}$$

Accordingly, as we can see from the structure notation, the two IIPs are now *related* through the function, and it is the function that as the object that is their relation, that indicates the two IIPs are distinct. This may seem very counterintuitive because in the more familiar description a function does not affect the axes, which remain orthogonal. We must remember that the IIPs *are not representations of axes*. We are in a different schema.

The IIPs represent the same fundamental thing that the axes represent - *quantity* - however, "axes" *always* represent it as numbers (which actually are number IIPs), whilst our IIPs here *may* represent numbers if that is what we specify $\{y\}$ and $\{x\}$ to be, but we may also, if we wish, specify that y and x are *continuum quantities*.

The usual schema using axes makes no distinction, but in the Object schema there is a distinction between numbers and continuum quantities. They are two distinct objects, even though real numbers can be used to express any continuum quantity. There are certain continuum quantities, however - the ones we must use irrationals to express - that real numbers can only express by virtue of the fact that they are IIPs.

We can see from the structure

$$\circlearrowleft (\infty) \{y\} \longleftrightarrow [\text{function}] \longleftrightarrow \circlearrowleft (\infty) \{x\}$$

that as long as this structure stands then the x and y IIPs are as inseparable from the function (which is a process object), as *it* is from the IIPs. It is the whole structure that is the object we represent as the corresponding graph line.

The structure

$$\circlearrowleft (\infty) \{y\} \xleftarrow{\quad \text{orthogonal} \quad} \rightarrow \circlearrowleft (\infty) \{x\}$$

is already inherent in the [function] object. So when we instantiate this structure all we are actually doing is instantiating a feature of the [function] object.

The orthogonality is inherent in the way the processes that are in the [function] object, already work. For example, if the [function] object is:

$$y = x^2$$

then as a structure that is:

$$\circlearrowleft (\infty) \{y\} \xleftarrow{\quad \text{must equal in quantity} \quad} \rightarrow \circlearrowleft (\infty) \{P(2) \circlearrowleft (\infty) \{x\}\}$$

This reads, literally, from left to right, "The state of the IIP for y quantities is always held equal to the that of the { IIP for raising a quantity to the power of 2, applied to to the state of the IIP for x quantities }".

Even if by "quantities" here, we assume we are talking about numbers, there is in the structure no intrinsic assumption that numbers are something that must exist independently of this structure, or that there is a structure

$$\circlearrowleft (\infty) \{y\} \leftarrow \frac{\text{orthogonal}}{\qquad} \rightarrow \circlearrowleft (\infty) \{x\}$$

that is first necessary.

These latter ideas, if they arise, are features of the way our intelligence is working, in understanding the nature of structures, rather than features of the way structures have to work, genuinely objectively.

In the case of a three-dimensional graph the structure is:

$$[\text{function}] \leftarrow \begin{cases} \rightarrow & \circlearrowleft (\infty) \{z\} \\ \rightarrow & \circlearrowleft (\infty) \{y\} \\ \rightarrow & \circlearrowleft (\infty) \{x\} \end{cases}$$

Here, similarly, this only seems to imply a "space" defined by x, y, and z, because of the way our intelligence is working in interpreting the structure. Any interpreting it in this way will come from the way we have been taught, which is, as it were, we could even say, a line of conditioning (literally in terms of brain function neuroplastic conditioning) that traces back to the ancient Greeks and to Euclidean geometry.

Nevertheless at its root, psychologically, the mechanism of this mentation arises from our experience of our world as sentient beings, whose experience of being and world arises through brain function. In other words, we experience a three-dimensional "space" between material objects, as something so fundamental to the way we experience the world, that in the way our mind works space itself becomes an object in its own right, for us. We then find we can measure space in its own right, increasing its credibility as an object, and we find that natural phenomena can be understood in terms of these measurements.

We find that the material phenomena is genuinely objective (by the criteria of Object Theory), and so from this situation, we somewhat unconsciously conclude that space must also be genuinely objective.

This situation persists to this day, right into the domain of relativity theory, where that theory's main object, which is *spacetime*, is taken to be genuinely objective. Or, we might say, *mistakenly* thought to be genuinely objective. When in fact, it is a mathematical construct that we use in order to comprehend the way in which natural phenomena takes place, on a scale greater than that in which we have our experience of self and world, as sentient beings, *which is all provided through the principle of brain function.*

In the Object Theory schema there happens to be no necessity of an object called the "space" that exists independently of the function, in order for the function to exist, and in which we then create the function, not withstanding that we currently use space as a conceptual tool for understanding.

Rather, the space we conceive (or indeed perceive) is *another object* that we construct independently from some of the same kind of objects that a structure of relations between quantities is composed from, namely, quantity or number IIPs that here we have labelled x, y, and z.

If we say that the relation in the structure of x and y quantities:

$$\mho\,(\infty)\,\{y\} \leftarrow \underline{\quad\text{orthogonal}\quad} \rightarrow \mho\,(\infty)\,\{x\}$$

is that the two quantities y an x are orthogonal, then we might have made a mathematical construct like an empty graph, and that is indeed a very familiar object in conceptualisation processes in mathematics. However, in itself, it is no more meaningful than the idea of infinite empty space and nothing else.

Such a conceptualisation is of course possible, and it may be that it will be assumed to be genuinely objective. But it is nonetheless still a chimerical product of the mind. The idea of space *is* a tool in current mathematics. Nevertheless, it arises not originally and in the first instance from a profoundly deep understanding of the nature of natural phenomena or numbers, as we experience them through the principle of brain function, but rather, from the experience of the world as a sentient being, whose experience arises through brain function.

A structure such as

$$\circlearrowleft (\infty) \{y\} \leftarrow \frac{\text{orthogonal}}{} \rightarrow \circlearrowleft (\infty) \{x\}$$

which is a mathematical space, even as a useful mathematical tool, only becomes more meaningful in the context of other structures into which it can be connected.

A structure such as

$$\circlearrowleft (\infty) \{y\} \longleftrightarrow [\text{function}] \longleftrightarrow \circlearrowleft (\infty) \{x\}$$

that we write and articulate in the more usual way, perhaps constructing a graph to show the function, is meaningful in isolation, in mathematics, as a way of showing how the quantities x and y behave with each other when they are *causally connected* through the *process object* that we call the function.

In mathematics we *causally connect* mathematical objects all the time. But we don't necessarily see it that way. In contemporary mathematics and science we tend to overlook this causal connection, or overlook that it is actually *causal*.

Mathematical causality is a fact, but it is not something we truly understand, because we tend in science to think of cause and effect as being something that only pertains to material interactions. And we also tend to think of our mathematical understanding of material phenomena as a "model", whilst thinking of the material phenomena itself as something other than just mathematical structure.

It is true that mathematical causality appears in mathematics in a way that seems detached from the actuality of the world, a way that is only theoretical. It is very different to how we experience what we call *cause and effect* in material phenomena. But now, also, we can see "pure" mathematical causality in material phenomena itself, in the behaviour of particles that are quantum entangled, whose states even when they are separated, are governed purely by their mathematical relations. We now know empirically as well as theoretically, that this is not due to any intervening cause and effect of material phenomena. Their entangled states are simply independent of time and space. If we wanted to, we might even see this as scientific evidence that time and space are not genuinely objective.

The Natural Continuum

We have been leading up to the idea of fractal structure without space. Most of our understanding of fractals is currently in terms of set theory. In these terms, a space consists of an infinite set of points. We then use the established ideas, such as the notion of hyperreal numbers, for how to relate an infinity of distinct points, to a continuum, in order to consider it as one of the same thing.

In Object Theory this doesn't hold, in the first instance, most simply, because a *continuum* is an object that is distinct from a "point" object, and as we have seen, there is no way of constructing a structure of *distinct point objects*, that is also a *continuum object*.

In an iteration function such as:

$$z \longrightarrow z^2 + c$$

which is the basis of the Mandelbrot set, z_0 (the initial value of z which in the Mandelbrot set is always 0) and c are always specific numbers on the complex plane. In the Mandelbrot set generation each infinite iteration of z therefore depends on c, which is chosen as a specific number on the complex plane. Hence, the iterations of z for all values of c other than 0, always create distinct z values even if the orbit is perfectly periodic and the value of z as the iteration continues revisits the same values of z infinitely many times.

Page 192

For example if we set $c = 1 + 0i$, the iteration is infinite instances of the number one. According to Object Theory, the reason these are *distinct*, is because they are *related* to each other by the iteration number.

If we want to have the complex plane as a continuum quantity that is not the same thing as an infinity of real, imaginary, and complex *numbers*, that we can then measure positions on, *using numbers*, then we are also going to be regarding the Mandelbrot set as a smooth function of the complex quantity continuum we call the plane, even though it is nowhere differentiable.

Earlier, we already gave one possible representation of the structure of a Julia set as a structure of *numbers* as:

$$\circlearrowleft (z_0) \, \{C\} \, \overbrace{\{\circlearrowleft (\infty) \, \{z \mapsto z^2\}\}}^{\uparrow | \checkmark}$$

Here, recall, to the left and outside the large horizontal brace, is a notation for an infinite iteration process of selecting numbers for the z_0 value (the initial z value in the iteration), that represent points the complex plane C.

We have seen that this process cannot cover the whole plane as a single IIP, even though it looks from our chosen notation as though it ought to be possible. We can just think of it as any process we like to imagine for selecting input numbers for the initial value of z in the iteration equation.

Underneath the large brace is the infinite iteration of the equation itself. Above the brace is the selection process for excluding from the connected Julia set those iterations that escape to infinity.

That is just a structural language representation of the well-known way of constructing a Julia set from complex numbers.

If we wanted to do the same thing for the Mandelbrot set, it would be:

$$\circlearrowleft (c) \, \{C\} \, \overbrace{\{\circlearrowleft (\infty) \, \{z \mapsto z^2 + c\}\}}^{\uparrow | \checkmark}$$

where now the iteration equation is

$$z \mapsto z^2 + c$$

The starting value of z is always zero, and now we are picking c values on the complex plane, using the process on the left, to put into the iteration equation.

So what happens if we don't use numbers?

It is normal to think of the complex plane as *consisting of* numbers. There is also much natural phenomena in physics and engineering, to say the least, that can only really be described and understood *using* complex numbers.

Both real numbers and imaginary numbers (and hence complex numbers) are things that have emerged from our own intelligence, as human beings. They are intimately linked with the way the natural phenomena of our world *is*. However, this is not because natural phenomena, when we are considering it objectively as something separate from our experience of it, is "made of" numbers, or "modelled on" numbers that belong to some transcendental reality separate from us.

Rather, the link between numbers and natural phenomena is in *us*. It is in the intelligence we are being, and in our brain function. The intelligence we are being arises out of natural phenomena. Part of its nature we encapsulate through the things we have invented (discovered in our own intelligence) and called numbers, together with the activity we have invented, or discovered in our own intelligence, called mathematics. It is naive to think that natural phenomena as we know it is our knowledge of something separate from our means of knowing it - which is brain function.

So if we allow that quantities in nature that we can measure, are not in themselves real *numbers*, other than the way we see them through our intelligence, and measure them, then we are also perfectly in order in saying that *complex quantities* are quantities in nature prior to our conception of numbers.

And so it is that the complex plane that we have constructed as a tool, represents a quantity continuum structure in nature, prior to our conception of it through the concept of numbers. In our notation we might try to write this natural continuum as:

$$\mho\left(\infty\right)\{A\} \longleftrightarrow \mho\left(\infty\right)\{I\}$$

where the object on the left is the real quantity continuum of whatever particular phenomenon it is we are measuring, and the object on the right is the imaginary quantity continuum for that phenomenon. However, because we cannot measure imaginary quantities directly, we don't know that the continuum in nature really has this two-part form. There is just something about the way quantities behave, the ones we can measure, that implies the imaginary part.

Imaginary quantities *are* a part of mathematical structures that *do* describe natural phenomena, even though we do not encounter imaginary quantities themselves, directly, in natural phenomena, whether as continuum or discrete quantities. We *must* therefore take imaginary quantities to be a mathematical representation of underlying processes and structures in natural phenomena that are inscrutable to us except through the artefact of imaginary numbers.

This leaves us in a position where complex numbers are essentially a tool we have invented (that is, discovered in our own intelligence) utilising the concept of $\sqrt{-1}$, in order to properly describe structures of continua in nature whose measurable properties can only be measured by real numbers.

In other words there is a continuum in nature that we can only describe and understand as two-dimensional number quantities, where the dimensions are the real and imaginary. The necessity for these two dimensions also arises in pure mathematics, but pure mathematics itself, only arises through natural phenomena, specifically, through human brain function.

When we describe quantity phenomena in nature with real numbers, what we are doing is creating the following structure:

$$\mho\left(R\right)\{Q\}$$

Page 195

which is a notation we have come across numerous times in preceding discussions. Essentially, here, we are applying infinite iteration processes

$$\circlearrowleft (R)$$

of generating real numbers, to the natural continuum quantity Q, of whatever kind of "substance" or "thing" that Q is a quantity of. Q is some quantity of some kind of phenomena, such as energy, or temperature, or length, that is distinct from other kinds.

So now we are looking at a deeper situation in which there is a quantity continuum object ♣ in nature, that is connected into structures whose natures determine the *qualitas* of whatever it is that exists as the continuum quantity are encountering.

In the consequent mathematical structures the measured quantity therefore becomes:

$$\circlearrowleft (C) \{♣\}$$

where C is the complex number we apply to the continuum, to make a measurement.

For a quantity continuum Q in nature, of some natural phenomenon, that we must measure with real numbers, we can write our mathematical representation of that natural quantity continuum, not just as Q, but also as the structure:

$$Q \equiv \check{Q} \longleftrightarrow ‡ \quad (1)$$

where now

$$\check{Q}$$

is an abstract mathematical *continuum* object, whilst ‡ is an object that represents the structure that identifies the *kind* of "substance" or "thing" that the phenomenon *is*, according to how we encounter it.

So if, for example, for Q we are talking about a quantity of *energy*, then ‡ is the structure that makes it *energy*, rather than say, for example, *time*. The nature of the structure will depend on how we are understanding say, *energy* or *time*.

The actual continuum in natural phenomena, ♣, is distinct from the *kind* of phenomenon, such as *energy* or *time*, that we encounter as a natural continuum quantity. Sometimes, the part of it that we have to represent with complex numbers, is present, and sometimes it is not.

It is not something that we can directly measure in natural phenomena. Rather, *it is present somewhere in the structure of* ‡, for any given type of phenomenon, whose quantities we can only measure with real numbers.

There is a structure:

$$\circlearrowleft (C) \{P\} \leftarrow \square \rightarrow \clubsuit$$

where $\square$ is the structure that relates and links (on the left) the application of numbers to the complex plane continuum through

$$\circlearrowleft (C) \{P\},$$

to (on the right) the natural continuum quantity ♣, in such a way that we can mathematically represent ♣.

The continuum ♣ itself always remains hidden somewhere in the structure of an encountered natural continuum ‡, the latter being the structure of a natural phenomenon *as we encounter it* in the measurable way.

The structures ♣ and ‡ are not yet things that we truly understand, in such a way that we know how they relate to brain function, which is the structure of phenomena through which, in the first instance, we encounter them.

Nevertheless, as we know, we don't *need* to understand that, in order to work out how ‡ *works* in nature *as we encounter it*, by mathematically modelling it in the models that we use in physics, for example.

This isn't, however, the end of the story for us, and it doesn't mean that we can just continue in this way, in the quest to find who we are and where we come from. If we are to bring our understanding of natural phenomena as ♣ and ‡ into the context of the understanding of brain function, through which all our understanding is arising in the first instance, then we must come to a deeper understanding of them, in their relation to structures in brain function.

We can regard continuum phenomena that we actually encounter in nature in a measurable way - *even when we only need to use real numbers to understand it with* - as part of a structure that at its core, has, or *is*, the ♣ object - the root object of continuum quantities in nature.

Of course, not all quantities arising in nature *are* continuum quantities. The most obvious example is that quantum mechanics deals with discrete quantities. Nevertheless, even though we may believe that quantum mechanics is fundamental to everything, continuum quantities do exist in nature. The relation between discontinuous quantities and continuous ones, cannot be simply explained away by saying the continuous quantities are really discontinuous, at a very small scale.

This also boils down to the question of the relation between numbers and the continuum. We try very hard to satisfy ourselves that distinct real or imaginary numbers can also be a continuum. Ultimately, though, we cannot realise our full potential of understanding, until we allow the continuum to be what it is, and relations of continuum quantities to be what they are, without the conviction that numbers come first. Then, we may be able to take the knowledge of numbers that natural quantity continua give rise to, through the principle of brain function, and apply them to structures in a new way.

We can ask that well worn question, does:

$$0.\dot{9} = 1 \ ?$$

This is such a common question. There are well known "proofs" that have been constructed to this effect. But there are also "proofs" to the opposite effect.

$0.\dot{9}$ is a geometric series whose common ratio is less than unity, so by the geometric series proof, the two numbers are equal. This glosses over the more fundamental issues that Object Theory deals with, but regardless of Object Theory, "proofs" in mathematics in the deepest analyses are only statements that are consistent with the axioms of the model. So which you want to go with, depends on your choice of axioms.

To complicate things more, there are in current mathematics, axioms that are undeclared, that an Object Theory approach would recognise. The fundamental undeclared axiom crops up as the mental action of taking certain objects of mathematical contemplation to be other than part of the structure of the mind, in such a way that a detached view of the underlying psychology of the reasoning itself is excluded from the picture, and objects are simply *presumed to be* genuinely objective (by the definition of Object Theory).

If we recognise that numbers are processes or parts thereof, then it is not difficult to see that the relation in

$$0.\dot{9} \longleftrightarrow 1$$

is not one of identicalness. However, the standard "proofs" that these two "numbers" are "equal", is about the quantities that they stand for.

In terms of Object Theory, the written number $0.\dot{9}$ (which is an IIP) stands for a quantity, which we can call

$$Q_1$$

Similarly, the written number 1 stands for a quantity which we can call

$$Q_2$$

and the assertion is that mathematically:

$$Q_1 = Q_2$$

which in Object Theory means Q_1 and Q_2 are two instantiations of one and the same quantity object, which we can call Q_3.

In the structure

$$0.\dot{9} \longleftrightarrow 1$$

both objects are IIPs, because in the same number construction system the 1 is also $1.\dot{0}$, also with infinite digits after the point. Both IIPs can only stand for the same quantity Q_3 *if* the rules of the IIPs allow this. This does not, however, mean that they *do*. We would have to *show* that they do. It would be a very specific kind of proof that we are not going to discuss here. (Regardless, $0.\dot{9}$ and 2 are clearly and obviously not two instances of the same symbol or number object, even if the quantities they stand for, are the same).

The arguments surrounding the question invariably arise from something more fundamental, and actually much simpler - the failure to acknowledge that a number and the quantity it is taken to stand for, are not two instances of one and the same object.

We need to aware of this in order to penetrate further into the question of quantity continua in nature, and their relations. A quantity continuum, being an object, has a structure. Its structure is not merely a structure of numbers. Its structure is a structure of IIPs, and is necessarily part of the structure of brain function, because it is only through brain function that any knowledge of quantity continua arises in the first instance.

Part 5
On the Mandelbrot and Julia sets

16: The Mandelbrot and Julia Sets as Structures

The form of the Mandelbrot set consists of a main cardioid, surrounded by *an infinity of bulbs*, plus what are often referred to as "filaments", that we shall elaborate on later.

The bulbs attached directly to the main cardioid we shall call "1st generation" bulbs. Those occurring around the circumference of the "1st generation" bulbs, we shall call "2nd generation" bulbs, and so on. There are infinitely many generations.

Each bulb has associated with it a *periodicity* of the orbit when c is within the bulb. This periodicity is, as we shall later see, linked to the structural form of the Julia set created when c is within that bulb.

Each bulb has a "basin of influence" of its periodicity. When c is inside a given bulb, but approaching its boundary, the structural form of the Julia set can be a superposition of influence from other bulbs.

As c crosses the set either from the main cardioid into a bulb, or from one bulb to another, the structural form of the associated Julia set undergoes a *transition* from one structural form to another.

In the transition from c values inside the first generation bulbs back into the main cardioid, the component parts of the structural components of the form of the corresponding Julia set overlap and merge with each other.

Later we shall see that there are two main IIPs producing the structural form of connected Julia sets: the SS IIP and the PS IIP. These are the Synapse Series IIP and the Parallel Series IIP respectively. In the transition of form that accompanies changes in c values within a first generation bulb to values within the main cardioid, the PS IIP becomes dominant as the only IIP producing the structural form of Julia sets. We'll talk about these later.

The larger the bulb, the more significant its influence into the surrounding part of the set. The crossing in the Mandelbrot set from one bulb to another of the next generation, has valleys either side, but there, the bulbs on the valley are so small that they have no discernible influence on the form of the Julia set, compared to the periodicities of the two large bulbs being crossed between.

The boundary between inside and outside the set, is always the boundary of a bulb, except:

- At the limit points of the valleys;
- At the boundaries of "filaments".

However:

As we described above, at or close to these limit points the form of the connected Julia set is uncomplicated by the influences of bulbs other than the bulbs whose edges are the valley. The smaller bulbs along the edges of the valley are vanishingly small by comparison and have little or no influence. The result is that the structural form of the Julia set is still characterised in the same way as it would be if c were well away from the limit point.

The same principle applies to the limit point of the valley on the right of the cardioid.

What are commonly called "filaments" of the Mandelbrot set are connected objects. The Julia sets arising from c values on computer-discoverable filaments or "mini Mandelbrots" are found to follow the same general structural principles of the SS IIP and the PS IIP that we are about to discuss.

In the following images the lighter shaded, contoured grey regions, are zones corresponding to the rate of escape to infinity of z, when z_0 is not in the Julia set.

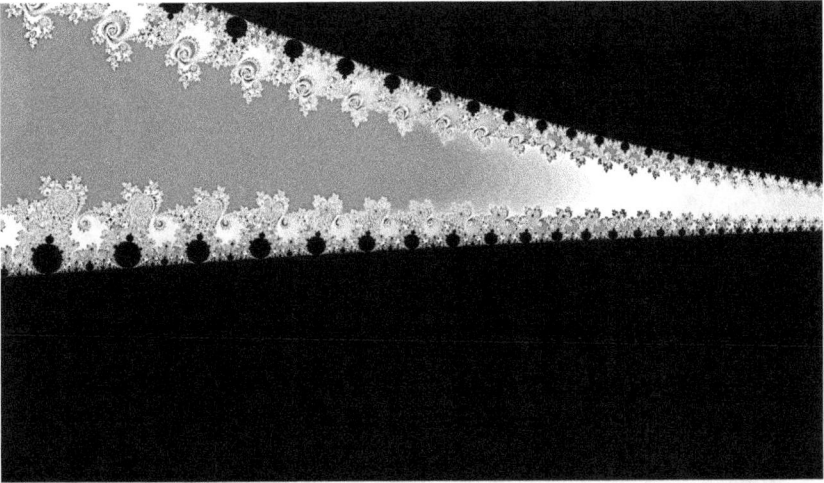

Above: The valley between the upper period 3 bulb and the cardioid, close to the valley limit.

Above: The corresponding Julia set when c is inside the central small bulb in the valley shown above. In the centre we see the initial symmetric, geometric object. The object connects at each end to a "synapse" object to which is connected many other objects. The number of objects connected into the synapse is equal to the period of the bulb.

Above: The corresponding Julia set when c is now outside the small bulb, just below it in the main cardioid. We can see how the overall form of the Julia set is the same, but the distinct geometric objects are now fused. Although c is inside the main cardioid, its proximity to the bulb still influences the form of the Julia set.

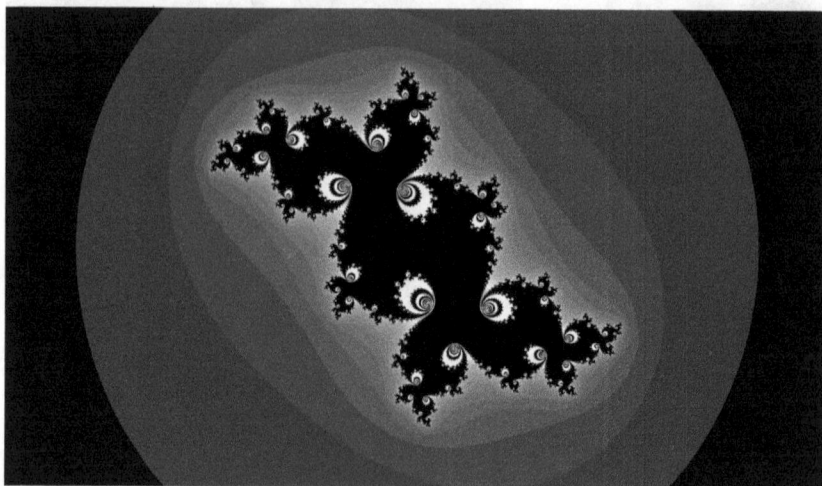

Above: The Julia set when c is now still within the main cardioid but closer to the valley limit. The influence from the bulb that is now closest to c is not significantly different to that of the original bulb, because all the bulbs in this vicinity have large period numbers and they do not differ greatly compared to the period number.

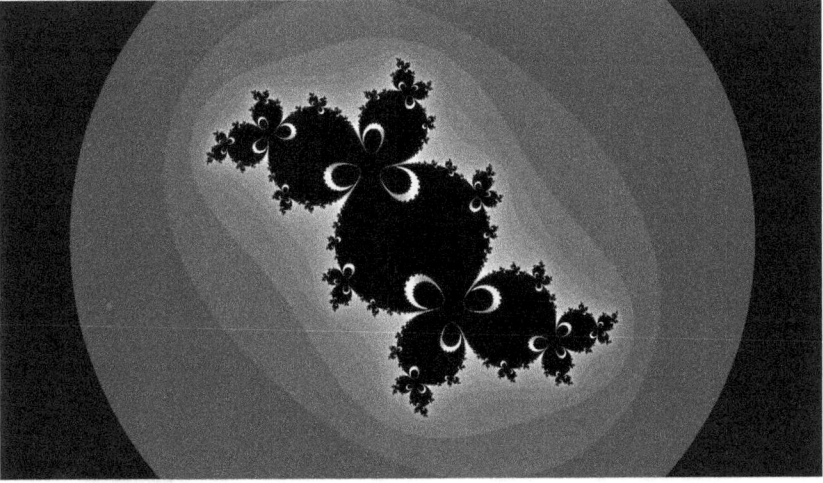

Above: Now c is very close indeed to the valley limit, and we can see how the form of the Julia set is beginning to show more prominently larger features that were incipiently present in the previous images. There is a central oblate circular shape attached to 4 smaller circular objects, two at each end. What we are seeing is the influence of the periodicity of the large period 3 bulb that c is approaching. This influenced can still be seen in all of the above images, but there, it is less pronounced. It becomes more pronounced the closer c is to the period 3 bulb.

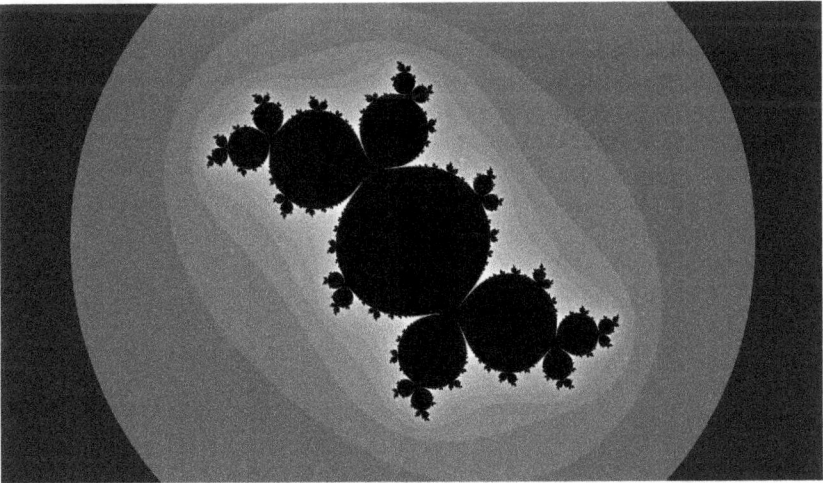

Above: c is now at the valley limit.

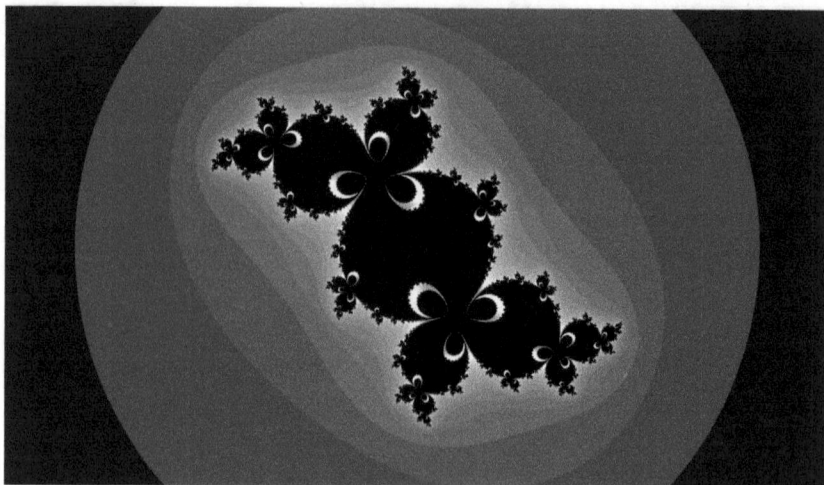

The previous reiterated for convenience. Above: Now c is very close indeed to the valley limit, and we can see how the form of the Julia set is beginning to show more prominently larger features that were incipiently present in the previous images. There is a central oblate circular shape attached to 4 smaller circular objects, two at each end. What we are seeing is the influence of the periodicity of the large period 3 bulb that c is approaching. This influenced can still be seen in all of the above images, but there, it is less pronounced. It becomes more pronounced the closer c is to the period 3 bulb.

Above: c is now at the valley limit.

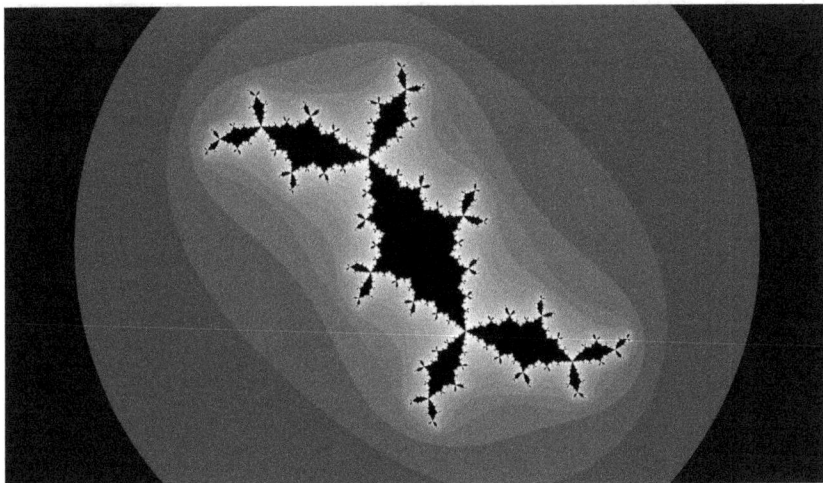

Above: *c* has now moved up towards the junction between the 3-period bulb, and the next biggest bulb attached to its edge, which is a second generation bulb, having the same main period of 3, but also a sub-period of 2. This means that *z* iterates in a main cycle of three positions, each position consists of 2 separate, close positions.

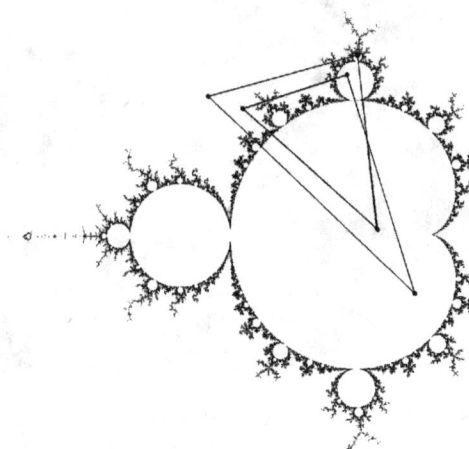

Above: The orbit of *z* when *c* is inside the second generation bulb. If we were to examine more closely we would find further sub-periods.

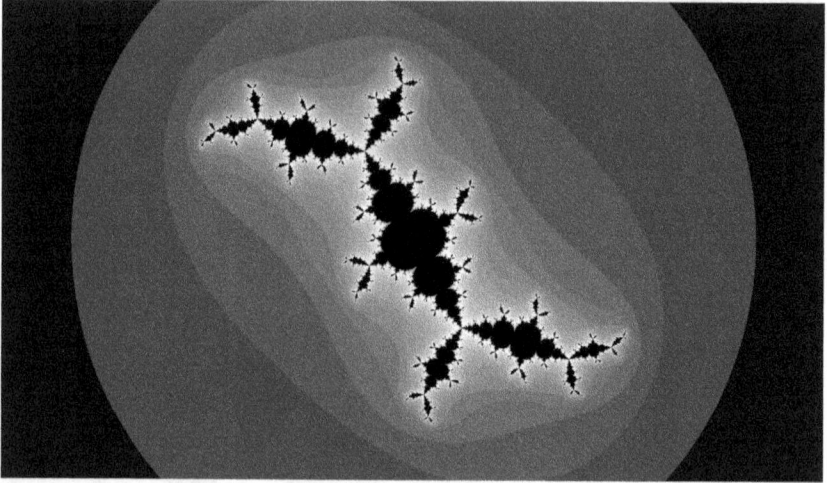

Above: c is now even closer to the second generation bulb. Note how the main components are now beginning to divide into separate components with an oblate circle-like shape.

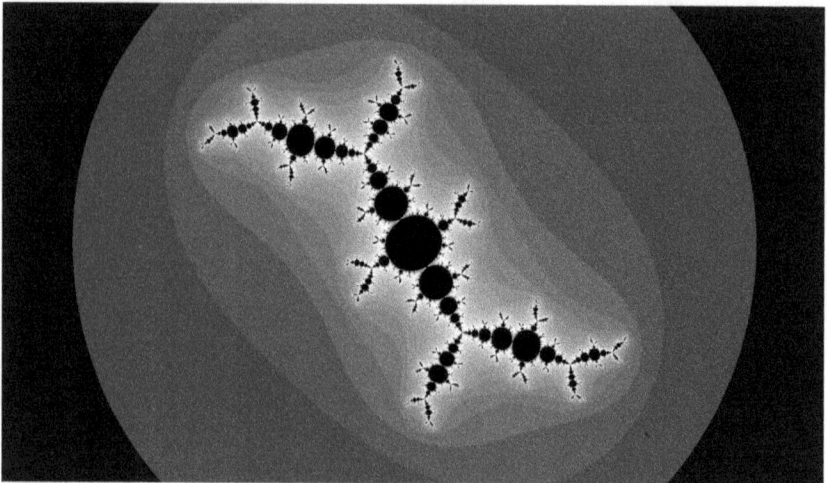

Above: c is now inside the second generation bulb. What was the main geometric shape objects, have now divided into further distinct objects - the oblate circle–like shapes. Note that the junction between these coming away from the origin, is a 2-synapse. This is the first sub-period number of the second generation bulb.

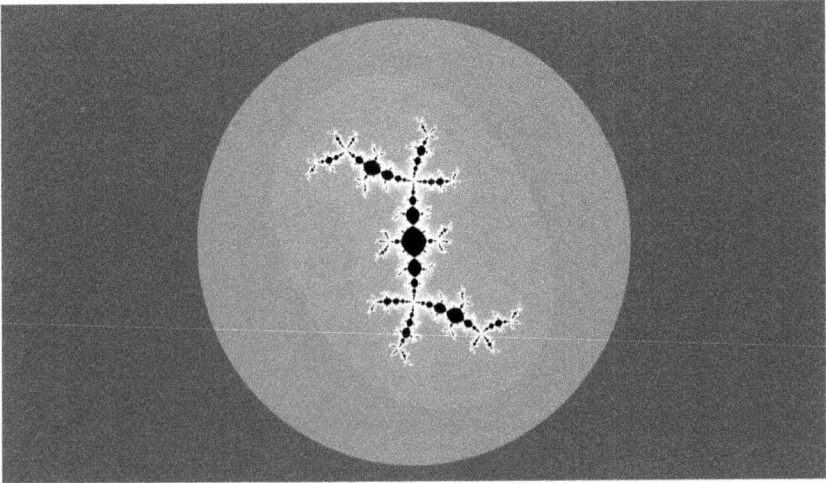

Above: The Julia set when c is in the centre of the largest 2nd-generation bulb on the edge of the upper 1st generation period 4 bulb of the Mandelbrot set. Note the 2-synapses in the series descending from the origin, and the 4-synapses further out.

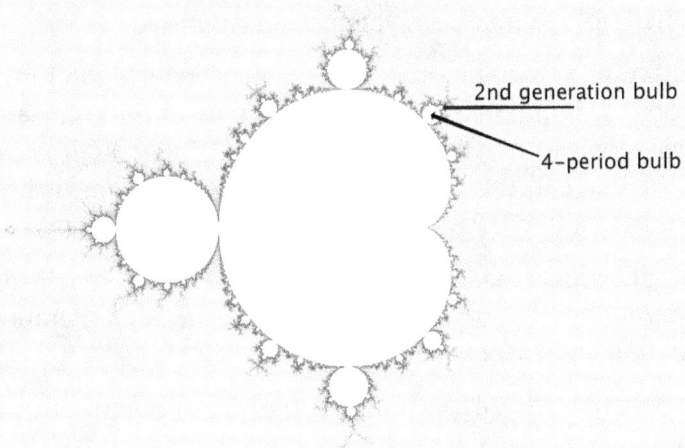

2nd generation bulb

4–period bulb

Above: The Mandelbrot set showing the position of the upper 4-period bulb, and the largest 2nd generation bulb attached to it, which has a period of 4 also, but a sub-period of 2.

The Synapse Series IIP

The form of a connected Julia set begins at the origin, where we find the initial symmetric, geometric object. When $c = 0$ this object is the well-known the filled-in circle of radius 1 whose centre is at the origin.

We see in action here, some of the rules that the progression of changes in the form of the Julia set follows, as c moves around the complex plane.

What we have been looking at is the progression of changes in the SS IIP - the *synapse series* IIP that forms the series of objects connected by synapses, extending outwards from the origin.

The SS IIP begins with the primary object, and extends outwards in two linear dimensions (lines that do not necessarily remain straight). It extends as a series of self-similar objects, through synapses whose order (the number of distinct objects connected to the synapse) depends on the position of c in the complex plane, and therefore, on the periodicity of the object in which c is positioned.

This is an infinite iteration process because it iterates the form of the primary object, infinitely. It creates objects as an infinite series that always *converges to a synapse*, and it is an IIP *that itself is infinitely iterated*. We can call this the *outer* SS IIP. In other words the SS IIP consists of two nested IIPs.

However, the inner and the outer SS IIP, may each apply a *transformation* to the object they produce on each iteration. The transformation has 4 components that are:

- Translation (change of position on the plane relative to previous iteration)

- Scaling

- Rotation

- Distortion

These four features of transformation are often utilised in iteration function systems (IFFs) that have been artificially created.

The transformations that take place allow morphisms in the form of the Julia set to take place in a smooth and orderly way as c moves around the Mandelbrot set, according to the "basins of influence" of the objects in the Mandelbrot set, in relation to the position of c.

Thus, as c moves across the complex plane *continuum*, the form of the Julia set in this respect (which is only one aspect of it) is a *structural function* of the complex plane continuum quantity that the number c represents.

The rule is that the inner SS IIP extends outwards along two *series axes* from the origin, in both directions along each axis. The series axis in each case is not necessarily a straight line with respect to the axes of the complex plane.

There is a *major series* in each direction along the longest axis, with larger objects than the minor series on the shorter axis. The minor axis is shorter because the objects along it are scaled smaller.

In all four direction the series converges to a synapse. At the synapse it meets all other series that are converging to that synapse. Thus the directions of series from the iteration of the primary object to the convergence at infinity on a synapse, alternate.

The SS IIP plays only one part in the formation of the Julia set. Another part is played by the PS IIP - the parallel series IIP. Let's look at some more images, illustrating the PS IIP in play.

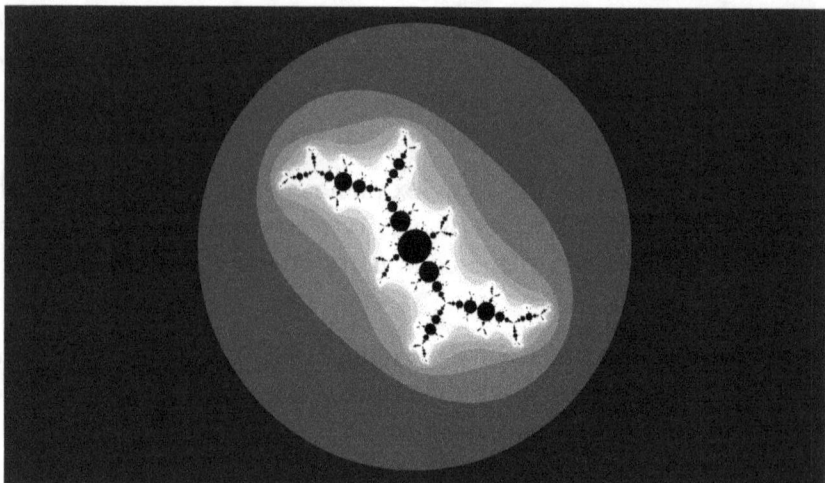

Above: The Julia set when c is in the biggest 2nd generation bulb of the upper 3-period bulb on the main cardioid. In the next image we will zoom in on the right-hand valley of the junction between the central primary object and large object just above it.

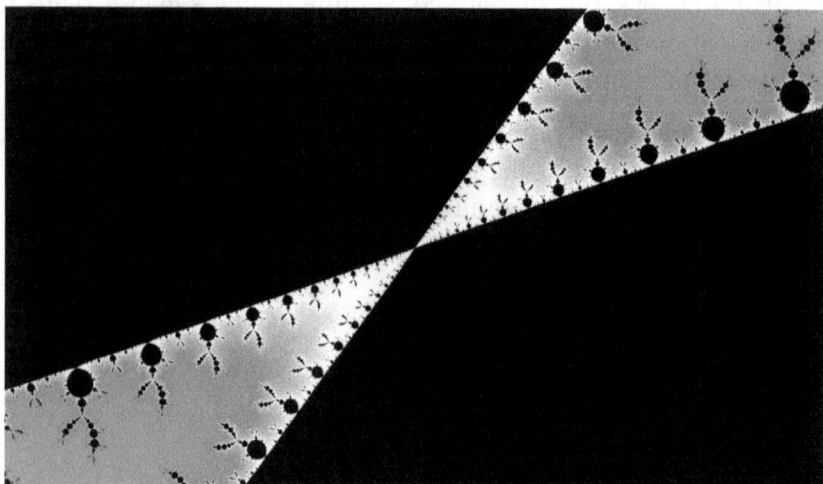

Above: Zoomed in on the valley described above, at X 500 x 10^6. Note the lines of "parallel" bulb objects similar to those found on the Mandelbrot set.

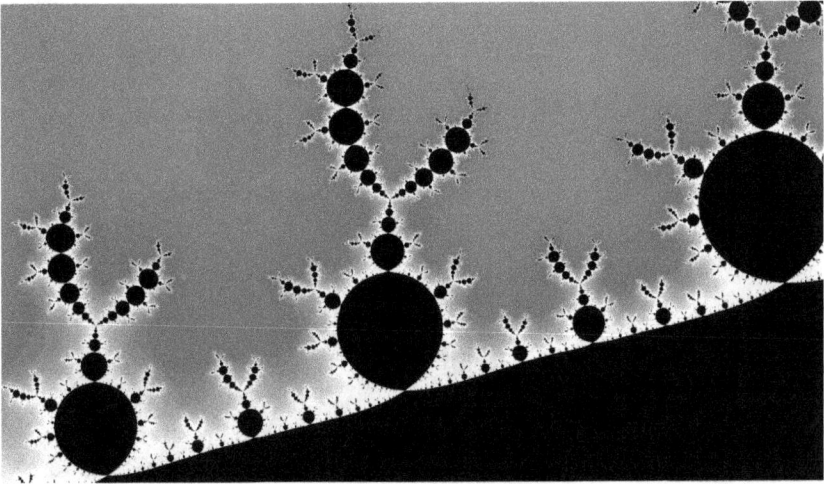

Above: Further up the valley, at zoom X 58254, more detail of the bulbs. What we are observing is an infinity of secondary SS IIPs "in parallel" along the edge of the object that is itself an object in one of the four primary SS IIPs. We can see that there are 3-synapses and 2-synapses, indicating the presence of those periodicities as main period and sub-period, in the bulb in which c is located in the Mandelbrot set.

It is important to appreciate that there are infinitely many bulbs, between the ones that are visible, and around the edge of every bulb, of any size, is similarly, infinitely many bulbs, to which the same principle then applies again, *ad infinitum*.

Thus, what we are looking at here, is an *infinity* of "parallel" series of *infinite nests* of infinities of "parallel" series.... *ad infinitum*, also with the objects always arranged according to the IS IIP principle that we will say much more about later.

These infinities *all converge* in one way or another. We have already seen that every instance of the inner SS IIP always converges on synapses. The outer SS IIP that instantiates the inner SS IIPs itself converges always on some point on the complex plane, and ultimately, on the boundary of the set.

In the next images, we'll see some more examples of how the parallel series IIP converges.

Above: We are going to look at part of the Julia set when *c* is in the white circle marked.

Above: The black area is the (filled in) Julia set. Typical of such fractal lines or edges or boundaries, we see parts that relatively *protrude* into the space beyond the set, and parts that *ingress* into the set object. It is relatively difficult to visually discern a "parallel" series of iterated objects like we saw in the previous images, but it is there.

Above: Now *c* has moved further towards the bulb, and the Julia set has undergone a corresponding morphism. Note that the parts of the boundaries that were previously ingresses, are becoming *valleys*. The previously protruding parts are now more protruding. On the left going down into the valley we can begin to see a "parallel" series.

Above: *c* has now moved further and the form of the Julia set shows more influence from the periodicity in its location. Clear valleys are showing, with "parallel" series.

Above: We have now zoomed in on one of the valleys in the image above. The "parallel" series of objects running along the edge of the valley are clearer.

Above: We have zoomed in more on the "parallel" series and can begin to see that on the main boundary large objects that are in a geometric series of diminishing size have smaller objects interspersed between them.

Above: Now we have zoomed in on a valley between one of the interspersed objects and the main boundary. We can see that the pattern of interspersing objects continues. Always between the largest object and the next size down, another object occurs that is the largest of all the objects between the first two. This occurs at any magnification factor (zoom level). Objects are arranged in geometric series of sizes according to this rule.

Above: In other locations, such as this one when c is close to "seahorse valley" between the main cardioid and the period 2 bulb on its left, the rule may be more difficult to discern, but it is there.

Above: Even on spiral paths the rule applies, but may be less obvious.

Above: The rule is present here, too. The valleys themselves have further valleys as part of their "edge", which complicates matters, but the rule still applies. The interspersed objects are "parallel" along the valley edges.

Above: One of the valleys on a valley side, from the area above, magnified. Some member objects of three different series are marked. "Parallel" series always occur on valley sides, or along the edge of a larger object.

Below: Non-valley "parallel" series formation in the Julia set with c in the main "minibrot" on the filament to the left of the Mandelbrot set.

The Parallel Series IIP

The parallel series IIP can be characterised in the following way:

We begin with two distinct instantiations of an object F, that are distinct because they appear subject to different transformations, that is, opposition, rotation, and scale. The transformation is what *relates* the two objects:

$$F \longleftrightarrow F$$

Each instantiation is uniquely configured, by applying a transformation T_n to the object F. So we could alternatively number the objects, and the transformations, and say:

$$F_1 \equiv \left[T_1 \overset{\longrightarrow}{\longleftrightarrow} F \right]$$
$$F_2 \equiv \left[T_2 \overset{\longrightarrow}{\longleftrightarrow} F \right]$$

where the direction arrow $\longrightarrow$ on the relation indicates the action of the transformation of the object F.

In the Julia sets the object in question is the geometric feature that begins as the primary object, and we see the transformation as a *change of location* on the complex plane, *scaling*, possible *rotation*, and possible *distortion*. So:

$$F_1 \longleftrightarrow F_2$$

would refer, for example, to two of the large black lozenge shaped objects that we can see in the last illustration above, protruding to the right and upwards, from each pinnacle on the fractal line from top left and bottom right. This is a parallel series of objects.

It is the transformation that makes these objects distinct. For example, they are distinct because they have different sizes, and because they appear in different positions on the complex plane. By the premises of

Object Theory the relation arrow is an object in its own right, and as such, we can say:

$$[F_1 \longleftrightarrow F_2] \equiv [F_1 \longleftrightarrow F_3 \longleftrightarrow F_2]$$

Abstractly, we can carry on replacing *relations* with specified objects, *ad infinitum*. This process is an IIP in its own right.

However, in the case of the parallel series IIP, if we want to model the geometric objects in the fractal, we need to be more specific. In the PS IIP a converging infinite series of objects is created by iteratively transforming an initial object.

This initial object is iteratively copied in this way, in an infinite series of instances of descending size, in consecutive positions, between the initial object, and the valleys of the two objects closest to the initial object that are larger than the initial object, to which it is not directly connected .

This principle is iterated infinitely with every parallel series object. The easiest way to see this is by looking at some examples.

Let's begin by going back to the last Julia set we partly illustrated above, and look at the form of the overall set, on the next page. Firstly, however, we must look again at the *synapse series* IIP again, because the *parallel series* IIP can only exist in the first instance because of the synapse series IIP, and, as we shall see, specifically because the *synapse series* IIP takes place on two axes.

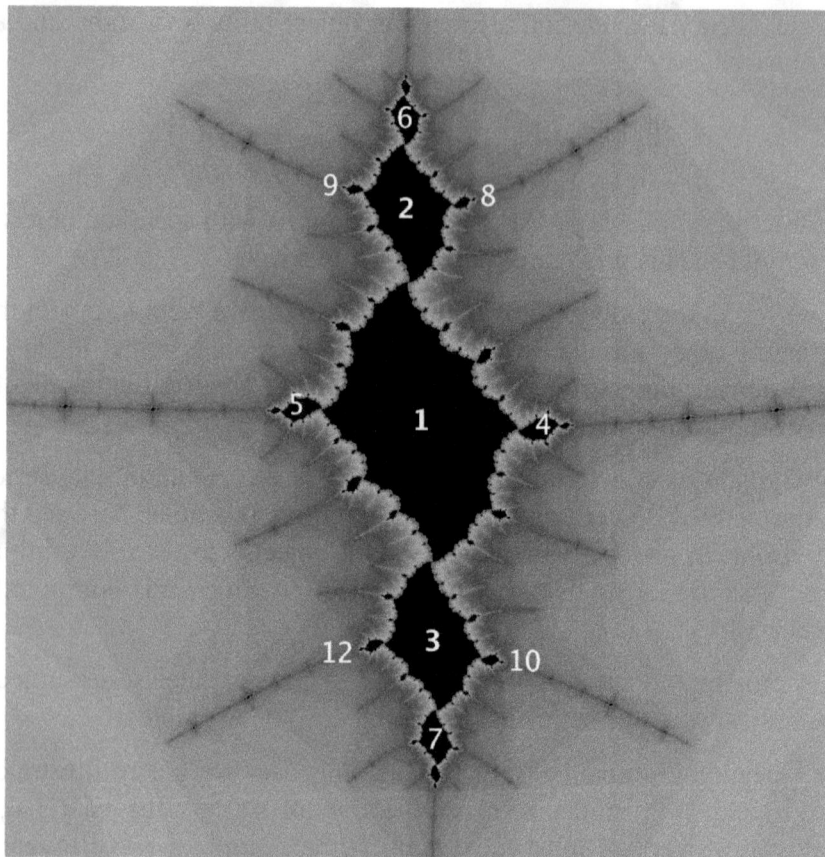

Above: The same Julia set as before, zoomed out. Object 1 is the central part of the form of the previously illustrated Julia set, before zooming. What we are looking at in the arrangement of the large diamond-like objects is a synapse series IIP (SS IIP) where everywhere the synapses are *2-synapses* (indicating c is in a 2-period zone).

The first iteration of the inner SS IIP is object 1. On the top and bottom and left and right corners are synapses. These are on the major and minor axes of the SS IIP instantiations.

Attached to the other side of the synapses, object 1 is instantiated again under transformation, 4 times, once on each synapse, as objects 2,3,4,5. This is the second iteration. Now these 4 objects have one synapse already attached to object 1. So the third iteration extends from the remaining 3 synapses on each object, as 9,6,8 and 12,7,11, and the corresponding much smaller objects attached to objects 5 and 4 (unlabelled).

The inner SS IIP instantiates infinite "copies" of the 1-object, along the major and minor axes, the re-instantiated objects always subject to *transformation* as:

- Change of position on the complex plane
- Scaling
- Possible rotation
- Possible distortion

The inner SS IIP may - depending on the Julia set - then be instantiated infinitely many times by the outer SS IIP, in an infinity of locations *other than* those that occur along the edges of the Julia set between the synapses. (The latter are accounted for by the *parallel series* IIP).

Essentially, the inner synapse series IIP is nested inside the outer SS ISP, which instantiates it (also subject to change) on each iteration, infinitely many times. We typically find these instantiations along "filaments" and where the inner SS IIP converges to a synapse, into which further instantiations of it are connected, the opposite way around.

The fact that the SS IIP axes in this example are vertical and horizontal is only because we have chosen a set in which this happens to be the case. In general, the axes can be rotated any amount, and furthermore, they are not necessarily even orthogonal to each other in terms of the complex plane. Nevertheless, all the Julia sets are symmetrical.

The Julia set we have chosen here is also relatively straightforward for the purposes of describing the IIPs through the geometry of the objects.

On the next page are two more Julia sets when c is in the period 2 bulb of the main part of the set. In this case, the axes are rotated approximately at right angles and the form of the set is much more complicated.

Above: The Julia set when c is in the position shown below.

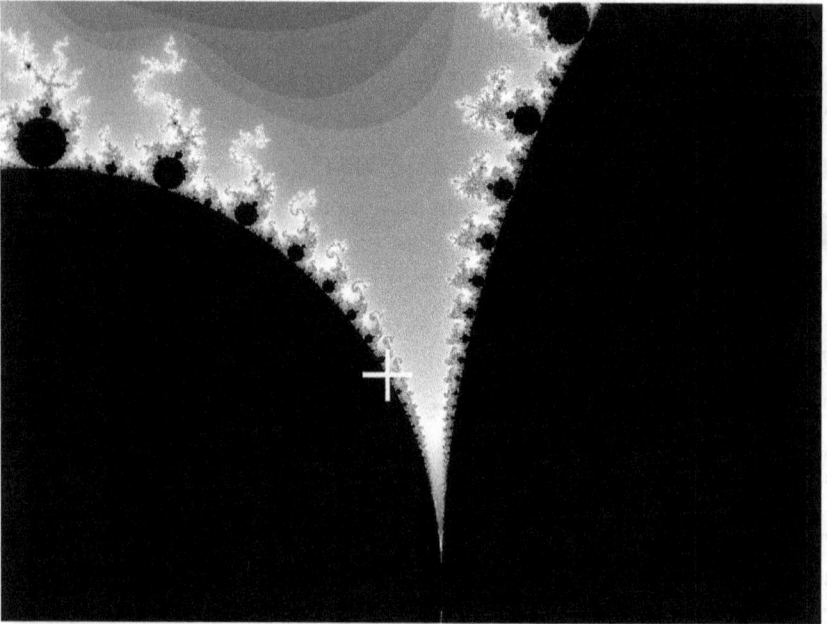

Above: The location of c producing the Julia set illustrated above. The location moves slightly closer to the boundary of the set to create the Julia set on the next page.

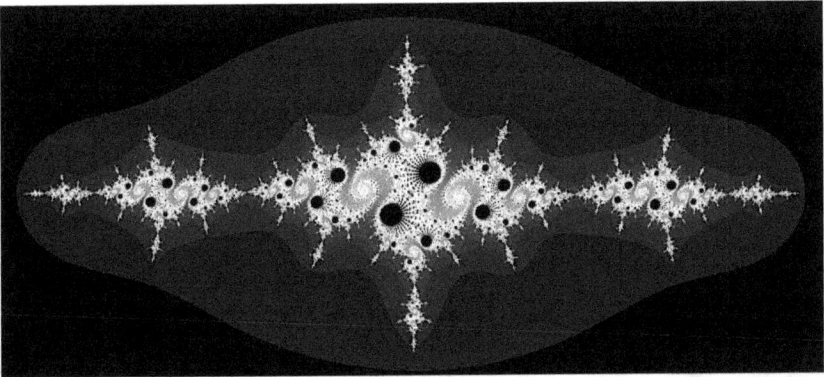

Above: An even more convoluted Julia set. c is slightly closer to the boundary of the Mandelbrot set, and the form of the Julia is under the influence of multiple periodicities.

Above: Closer zoom on the central part of the above set. Computer generated imagery can be misleading because the resolution is limited by the number of iteration used in the algorithm. In these images, what may appear to be large synapses within the set, are in fact unresolved areas consisting of areas both inside and outside the set. Nonetheless, we can still see the principle of the SS IIP operating in the same way.

Note how in the whole "zoomed out" set there is the SS IIP series of objects extending away from the origin in the same way as before, and each large object consists of an SS IIP series in its own right.

Spiral Formations

The form of *these* Julia sets now shows more convolution, and in particular, the spiral formations that are often seen in Julia sets. Spiral formations are a spiralled SS IIP series of objects, converging to a synapse. Whenever this happens valleys are formed between the arms of the spiral, as well as between the objects that form the arms of the spiral.

In the case of spirals, the valley between the arms is itself spiralled. Along the sides of the valleys there is a parallel series IIP formation that may be indistinguishable from the synapse series formation. They can essentially become the same thing.

Two Axes of the SS IIP

The *parallel series* IIP can be thought of as initiated in terms of the *axes* of the *synapse series* IIP. Because the synapse series IIP takes place along two axes, the major and minor, there is inevitably a *boundary* of the Julia set between one axis and the other, that occurs on the primary object. This then occurs in any subsequent iterations of that object by the synapse series IIP.

Thus parallel series IIP formations always occur along a line that leads to a valley, which is the connection of one object with another at a synapse, as part of a synapse series IIP formation.

The PS IIP and the SS IIP are inseparable, and interrelated, and geometric features must show elements of both.

The Parallel Series IIP Principle

The parallel series IIP follows the same principle everywhere in the connected sets, including in the Mandelbrot set. The principle is that any object is infinitely iterated (under transformation) in two directions away from the object, in a converging infinite series. The convergence always occurs at the limit point of a valley, or a synapse. Between the iterated objects of a converging series, in any direction, is always an infinity of further converging parallel series.

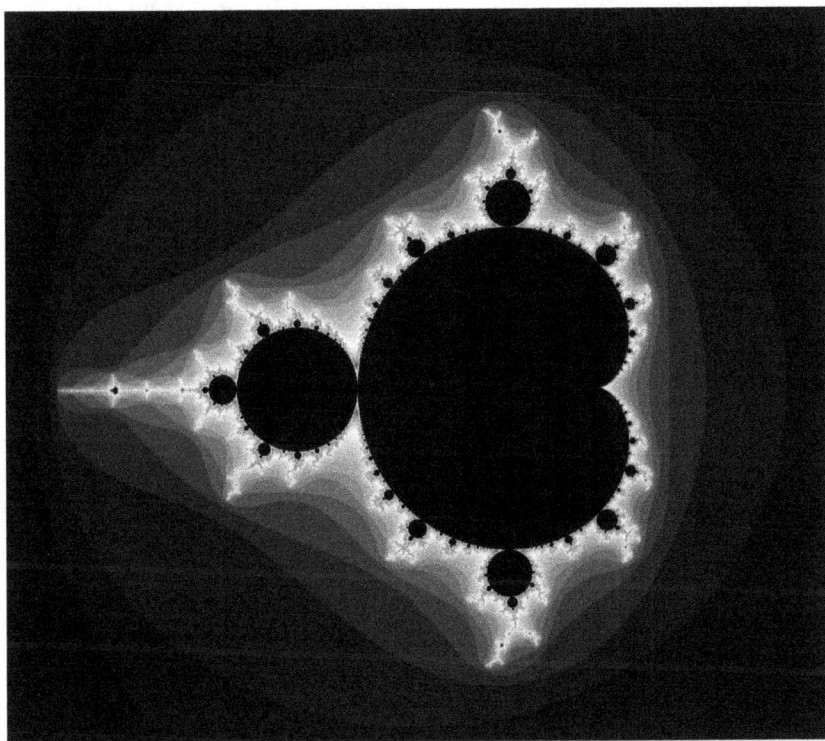

The rule is: A converging "parallel" series of iterated objects goes in the direction of convergence from an initial object to the valleys of the two objects closest to the initial object, that are larger than itself, one of which is not directly connected to itself. This applies to both the connected Julia sets and the Mandelbrot set.

However, true synapses in the Julia sets only occur where c is outside the main cardioid of the Mandelbrot set. Both valley limit points and synapses are convergence points of the parallel series IIP, and synapses always also consist of valleys converging. However, not all valley limits are synapses. The valley limit on the right of main cardioid of the Mandelbrot set is not a synapse.

One can see that the form of the Mandelbrot set itself complies with the parallel series IIP principle. It makes a good object of study for the relation between the PS IIP and the SS IIP.

We can take the entire Mandelbrot set object to be instantiated by the IIP that produces it, that we wrote earlier as:

$$\overset{\uparrow|\checkmark}{\overline{[\circlearrowleft(\infty)\{z\mapsto z^2+c\}]}}$$

However, the object itself as it appears geometrically exhibits the characteristic behaviours of the PS IIP and the SS IIP. Just as the iteration equation IIP that produces the object is not a process that is a function of time, so also the PS IIP and the SS IIP are not functions of time. So when we speak about objects being instantiated on a particular iteration of the PS IIP or SS IIP, we are not implying that the PS IIP and the SS IIP produce the Mandelbrot set object in stages, any more than the iteration equation IIP above produces the object produces at one point at a time.

Rather, it is the case that there is a structural, functional relation between the iteration numbers, and the object. Hence, there is a structural, functional *relation* between *the natural numbers* and the Mandelbrot set object (a relation that is the IIP) since the iteration numbers are nothing other than the natural numbers.

In order to appreciate the nature of the structural IIPs let's provide the image of the Mandelbrot set again, together with a commentary on how component objects might be viewed as being instantiated, *if it were the case* that the PS IIP and the SS IIP did instantiate these objects one object at a time.

The illustration and the commentary appears on the next page.

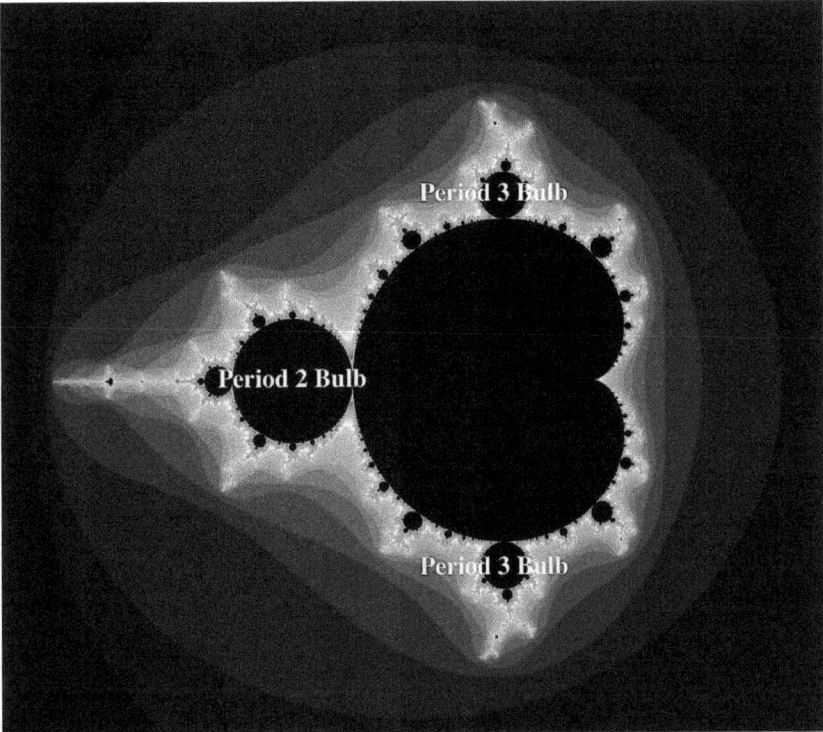

The instantiation would begin with the first object, which is the main cardioid. The next iteration is the instantiation of the large period 2 bulb to the left. Immediately we now have 2 objects, joined by a synapse and 2 valleys. So each side the large period 2 bulb a converging series of bulbs is iterated (each iteration subject to transformation), starting with the period 3 bulbs top and bottom of the main cardioid, and converging to the valley on the right of the cardioid. Next, converging series are instantiated left of the period 3 bulbs top and bottom, converging to the valleys at the synapse between the big period 2 bulb on the left and the main cardioid. We can see this reflected in the fact that the periods and sizes and positions of the bulbs on the left side of the cardioid are different to those on the right. There is symmetry top and bottom of the cardioid, but not left and right.

Next, we go to the next size bulbs down, that are part of the previous series. These are at positions intermediate between the big period 2 bulb and the period 3 bulbs. Converging series are then applied from these, converging on the valleys in the synapses between the cardioid and the period 2 and period 3 bulbs. And so on.

Following on from the commentary with the previous illustration, the bulbs already created become starting objects for a converging series of copies (subject to transformation) either side of each bulb, converging on the corresponding synapses.

The series instantiations and position of the bulbs are related to the *sizes* of the bulbs. The initial series from the big period 2 bulb on the left of the cardioid, starts with the period 3 bulbs top and bottom, and converges in the valley on the right of the cardioid.

As we have said, starting from a given initial object O there are two parallel series extending from it, one each side, that always converge at the valleys of the two objects closest to O that are *larger* than O.

This principle is the same for any connected Julia set, and applies as a rule for the placement of "parallel" objects, whether we are looking at singular objects or objects that are also combinations of multiple objects such as those that occur like arms around synapses.

Returning to the image of the Mandelbrot set, at the same time as the *parallel series* formations, the *synapse series* IIP places further bulbs to the left of the period 2 bulb, and further bulbs on the outside of the bulbs already instantiated in the parallel series. Further parallel series are then applied to these. And so on. The two IIPs work together to create the entire set.

There are situations, however, where objects descend in size and then increase in size again, without descending into a valley or synapse. An example of this is given in the illustration on the next page. It is not that the SS IIP and PS IIP are not present, or that there is some other IIP in action. Rather, it arises from the way the two principles of the AA IIP and PS IIP combine.

We see some illustrations of this on the next page.

Above: Detail from a highly zoomed part of the Julia set when c close to the edge of the Mandelbrot set big period 2 bulb. There appears to be a series of objects marked within the white ellipse that partially converges to a point in the middle that is not a synapse or valley.

Below: The central part of the same Julia set zoomed out, where we can see an earlier iteration of the same feature (rotated anticlockwise slightly) along the upper right-hand edge of the central primary object.

When such situations arise they are indicative of the conjoining of component objects, which *also always* happens when c is inside the main cardioid.

Above, and below zoomed in: When c is moved closer to the edge of the bulb the component objects begin to separate out. The sizes of the objects in question are a consequence of the SS IIP, rather than the PS IIP, and it is the SS IIP that iterates objects that converge into the synapse in the spiral.

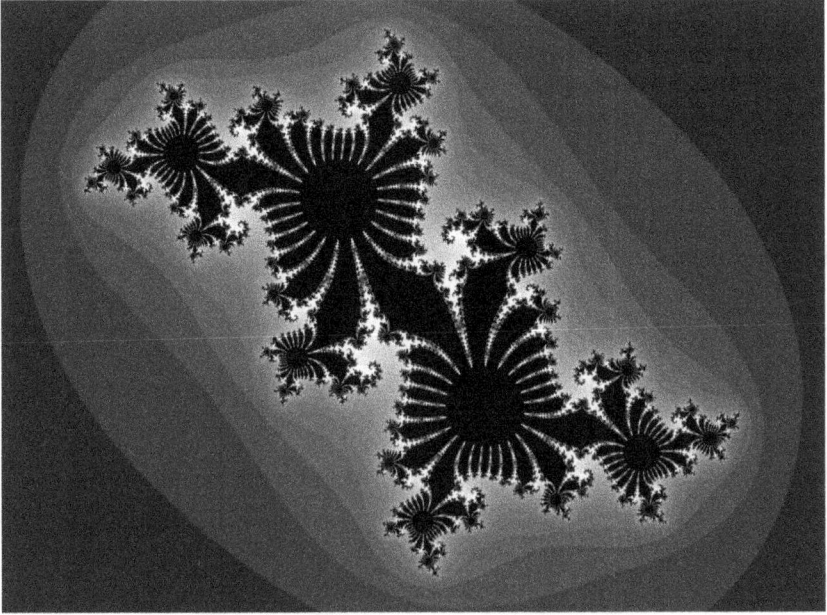

Above: Another example where c is in a small bulb close to the valley between the big period 3 bulb and the cardioid in the Mandelbrot set.

Note that the large black circle like areas are where there is insufficiently great iteration resolution in generating the computer image. With greater resolution we would see the gaps between the arms around the synapses descending right into the centre of each synapse position.

Again we can see that the SS IIP produces a similar effect due to the action of the translations that take place between one instantiation of the object and the next.

The Julia sets do not separate out into distinct objects until c is very close to the boundary of the main cardioid. Even in the objects outside the main cardioid the principle that there is more convolution, and more distinctions between objects, as she approaches the edge of an object, still applies.

When c is outside the Mandelbrot set, all Julia sets found in this region are *disconnected*. The objects and patterns observed in computer imagery of this region represent fractal zones of *rates of escape to infinity* of the orbit of z.

Summary - SS IIP & PS IIP

Synapse Series

When c is outside the main cardioid of the Mandelbrot set then the *synapse series* IIP iterates copies of the object under transformation, each in a converging series. The synapses are the joining areas or points of iterated copies.

The synapses define distinct objects that are nonetheless connected. The synapse series of objects converges to a convergent point.

At the convergence point at least one new synapse series of objects emerges (converges into). The total number of synapse series emerging from (i.e. converging into) a synapse, equals the *synapse number*. The synapse number is the period or sub-period of the bulb in which c is located, depending on the synapse.

When c is inside the main cardioid then synapse series do not clearly begin to form unless c is very close to the boundary between the cardioid and the bulbs.

When c is inside a "minibrot" cardioid, then the form of the Julia sets depends on c's position relative to the "minibrot" form, in a way exactly similar to how Julia set forms behave when c is in the main cardioid of the set.

However, when c is inside a "minibrot" cardioid then the primary object of the Julia set has the same form (although scaled and rotated) as the corresponding entire set when the c is inside the main cardioid.

There are also infinitely many synapse series of this form around the primary object of the Julia set, that *converge into* the primary object. Series of small iterations of the primary object lead to infinitely many large iterations of the primary object. The same thing applies to Julia sets when c is inside the build of a "minibrot".

On the following page are some examples of such Julia sets.

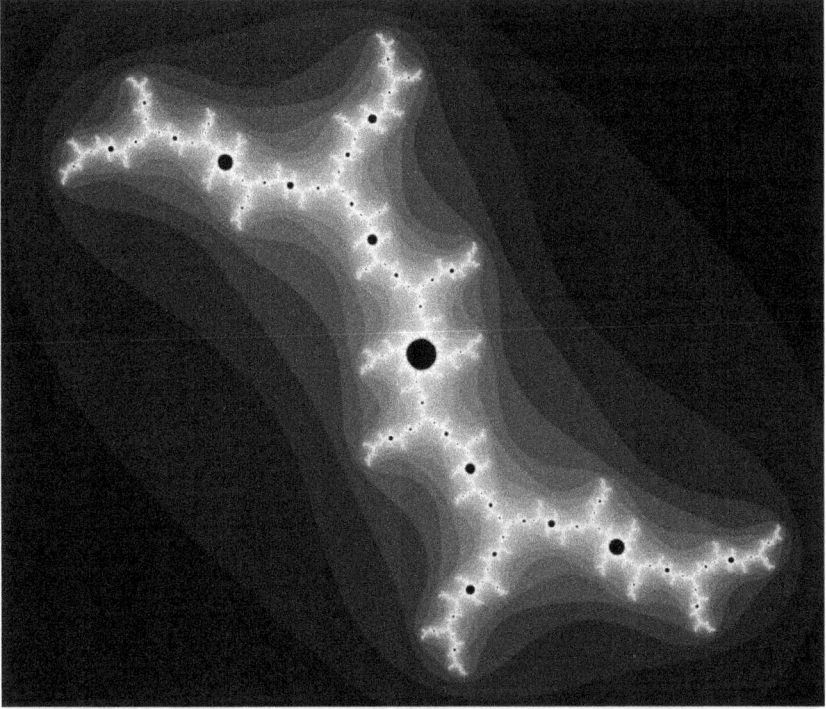
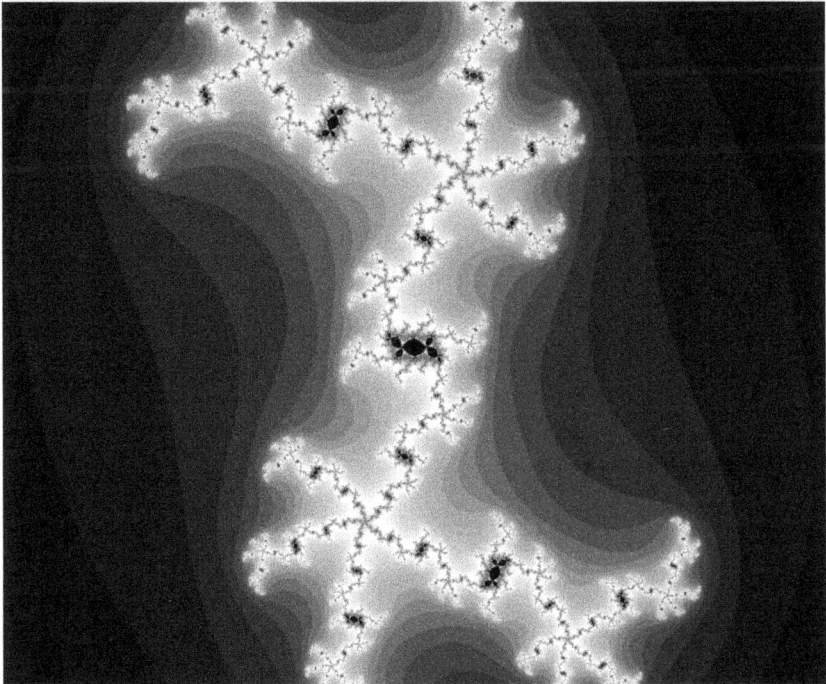

Page 237

Parallel Series

The parallel series iterates as a converging series of objects attached to an edge of a larger object. The rule is: A converging "parallel" series of iterated objects goes in a direction of convergence from an initial object to the valleys of the two objects closest to the initial

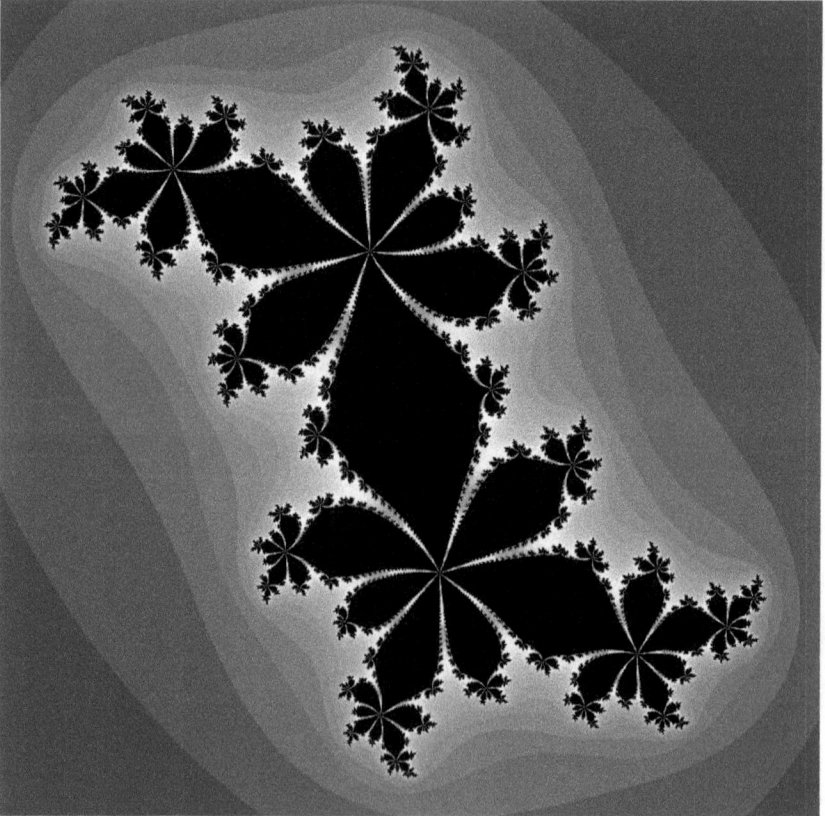

Above: Julia set in which the combined effects of the SS IIP and the PS IIP are easy to see as distinct processes.

We can see that from the synapses that c must be in a 7-period bulb. The PS IIP formations into valleys is easy to discern. We can also see from the absence of spirals that c must be aligned with the axis of symmetry of the bulb.

object, that are larger than itself, one of which is not directly connected to itself.

Combination

The two IIP types, the SS IIP and PS IIP, are never exclusive. The *parallel series* IIP only establishes its rule in conjunction with the *synapse series* establishing *its* rule. This interdependence is most obvious in the way Julia set forms behave for c inside the main cardioid. Where conjoining of objects occurs, it may not be obvious which IIP is dominant.

17: Conjunction, Connectedness and Disconnection

By definition, the *connected Julia sets* are connected, meaning that whilst the geometric objects we have been talking about are distinct objects, they are nonetheless connected. It is perfectly possible for objects to be distinct, and yet connected. In terms of Object Theory, this just means that in the relation

$$F \longleftrightarrow F$$

where the Fs are two distinct instances of an object F, this object F is something that *does not require* for its definition that it is separate from other instances of itself. This means that in the structure

$$F \longleftrightarrow F$$

the *relation* is not one of separateness. Rather, as in the case of the fractal forms it is one of *similarity*.

For convenience and compactness, we are now going to change the notation we have been using. Firstly, let us symbolise an IIP such as the ones that involves geometric objects, as:

$$\odot$$

Standing by itself, this symbolises an "empty" infinite iteration process, with no process specified. It is simply the *infinite iteration object*.

Now we start with the SS IIP that proceeds along two iteration axes, and show that the IIP is happening simultaneously along the two axes as:

$$\oplus$$

The IIP proceeds along the two axes, branching and spreading through synapses according to the periods and sub periods in the location of c. So this IIP just depends on c. So we can symbolise the "empty" process so far as:

$$\oplus [\odot_c]$$

Applied to the fractal, it is a process of iterating transformed copies of the primary object P. This includes, where necessary, making n copies at some synapses, where n is the period of the first generation bulb). Any other bulb that c is in, is a "descendent" of this first generation bulb, whose periodicity continues to have an effect.

A transformation array T applied to P would be T : P. However, each transformation T depends on the individual iteration of the SS IIP. So the process is symbolised:

$$\oplus [\odot_c (T_\odot : P)]$$

where $T_\odot$ is a transformation IIP in its own right, whose iterations are the iterations of $\odot_c$. $T_\odot$ constitutes an infinite series of transformation specifications that may differ in 4 directions, 2 directions on each iteration axis, on paths towards and away from synapses. The precise specifications depend on the individual iteration of $\odot_c$, as indicated by T's subscript.

This completes the synapse series IIP.

The parallel series IIP is carried out descending into every valley edge of every object in the synapse series IIP construct. The transformation applied at each iteration of the parallel series IIP again depends on c and may be different for each valley edge of an object. It also depends on each iteration of $\odot_c$. So this multidimensional transformation V is:

$$V_\odot$$

and it applies as a transformation to every object that the SS IIP instantiates, on each iteration. It instantiates the infinite series of objects on the edges of the transformed instantiations of the primary object made by the SS IIP (which includes the primary object as its iteration 1).

So the parallel series IIP is:

$$\odot_c V_\odot : \{\oplus[\odot_c(T_\odot : P)]\}$$

How is this? Here, starting on the left, the infinite iteration $\odot_c$ has a subscript c to show that what happens *within each of its iterations* depends on c.

The subscript to V shows that V is dependent on each particular iteration of $\odot_c$.

So the combined term on the left of the colon is stating that $\odot_c$ iterates infinite instantiations of the transformation $V_\odot$, which is itself an infinite array. Following the colon, on the right, is *what this transformation applies to*. So $V_\odot$ transforms the object in the curly brackets on each iteration of $\odot_c$.

The object in the curly brackets is what we have already shown is the infinity of geometric objects that the synapse series IIP instantiates. However, here, this is *not* the instantiation of this object made by the SS IIP. Rather, it is a second instantiation made by the PS IIP itself.

So the transformation $V_\odot$ is being applied to *every geometric object* instantiated by the object:

$$\oplus[\odot_c(T_\odot : P)]$$

In other words, there is no geometric object instantiated by the PS IIP that is not also an object that has been instantiated by the SS IIP.

Both IIPs instantiate each and every geometric object. The geometric objects only come into existence through the *combination of both* IIPs.

So one way of representing the entire fractal geometric object of the connected Julia set, and indeed the boundary of the Mandelbrot set, is as a *structure* $\mathfrak{M}$:

$$\oplus[\odot_c(T_\odot : P)] \longleftrightarrow \left[\odot_c V_\odot : \{\oplus[\odot_c(T_\odot : P)]\}\right]$$

in which the two instantiations, left and right, of the object

$$\oplus[\odot_c(T_\odot : P)]$$

are not representing two distinct objects, but are just two distinct instantiations of the same object, the synapse series IIP, in this structure. This structure lies behind all appearances of connected Julia sets, and is a structure that in itself as an object structure (as distinct from the actual geometric forms it *produces*), is invariant with respect to c.

Let us consider the Mandelbrot and Julia sets when c values are outside the main cardioid. Here, c is *always* within a bulb, unless it is considered on the synapse between two bulbs. There is nowhere else for c to be as long as it is within the Mandelbrot set.

There are infinitely many infinities within infinities of parallel series of bulbs around the boundary of the cardioid, and there is no "gap" between bulbs. Rather, the existence of the bulbs is an infinitely nested IIP that itself *constitutes* the boundary of the cardioid. This is the IIP that we describe with the structure given above.

Some salient points are as follows:

- Essentially, the parallel series IIPs is exist as an infinity of synapse series IIPs arranging themselves "parallel" to themselves, infinitely many times, and at every opportunity.

- However, just as the synapse series itself must follow a transformation rule that always leads to a convergence point, so must this "parallel" arrangement. All series of objects converge to finite convergence points on the complex plane.

- With the *apparent* exception of the valley on the right of the main cardioid (which we will come back to) of the Mandelbrot set, the limit point of all valleys are synapses. So *all* the limit points, are *synapses*.

- The series of transformations of both the synapse series IIP, and the parallel series IIP, always converge on a synapse.

- Every *parallel series* consists of objects that are in a *synapse series* that includes a common object to which all the other objects in the parallel series are directly connected.

In the connected Julia sets:

- Every synapse is a convergence point between n objects (determined by the periodicity and sub-periodicity of the bulb at the location of c), and $2n$ parallel series.

- Each geometric object in each parallel series is itself a *complete synapse series* or set of synapse series with *the same structure* as the synapse series on which the parallel series occurs.

Structure

The point about the structure, is that it is always the same, even though the geometric manifestation of it differs in different locations, due to the action of the transformation objects that are contained as part of the structure.

In the connected Julia sets there is *one* overall structure, that is iterated infinitely many times, under transformation, one infinite iteration on the major axis, and one infinite iteration on the minor axis.

In the Mandelbrot set all the second and subsequent generation bulbs have a first sub period of 2, and in the connected Julia sets, from the origin, the initial synapse series extends from the origin out through both axes, through 2-synapses (synapses between two synapse series).

Conjunction

When c is within the main cardioid of the Mandelbrot set the distinct geometric objects that we see in the Julia sets for c values within a bulb, become conjoined. There are then no true synapses, but rather, the synapses become areas of *conjunction*.

When c is close to the origin, distinguishable objects only occur as fractal features of the boundary of the "filled in" Julia set. Total conjunction between all objects occurs when $c = 0$, and the "filled in" Julia set becomes a circle around the origin of radius 1.

When c $\neq$ 0 but is still well away from the boundary of the Mandelbrot set there appears to be no synapse series present, and any parallel series of boundary features appears not to converge to a point on the plane. However the principle of both series is still implicit in the form, determining the form of the features, but only emerges fully when c moves outwards along a radial from the origin to *inside* a bulb.

The connection of any bulb to the main cardioid is a synapse on a diameter of the bulb, which the radial must intersect. The radial relates to the orientation of the major axis of the Julia set. The degree of conjunction between the geometric objects in the Julia set decreases by some inverse function with the distance of c from the origin. All conjunction areas become synaptic points of connectedness, once c is well within a bulb.

When c = 0 the geometric form of the Julia set is totally conjoined, as the "filled in" circle around the origin of radius 1. In the structure we found:

$$\oplus[\odot_c(T_\odot : P)] \longleftrightarrow \left[\odot_c V_\odot : \{\oplus[\odot_c(T_\odot : P)]\} \right]$$

how is this represented?

The primary object P is now the "filled in" circle. The axes $\oplus$ are now the complex plane axes. The infinite iteration $\odot_c$ still exists, which means that the primary object P is still being instantiated infinitely many times, still on each axis, unchanged.

As c moves towards 0, the conjoined areas arising from the synapses increase in size, and the scaled, iterated objects of the SS IIP also increase in size, the conjoined areas increasing. The The smallest PS IIP objects merge soonest as c decreases, the largest ones latest, all becoming absorbed into the conjoined area at $c = 0$.

At $c = 0$ the parallel series IIP on the right therefore now consists of the transformation $V_\odot$ being applied infinitely many times to what is now the primary object P, the "filled in" circle.

However, as we know, at $c = 0$ there are no actual transformations to the primary object that appear in the set, so the synapse series $T_\odot$ and parallel series $V_\odot$ transformation objects are having no effect. Nonetheless, as we also know, the "filled in" circle is still the consequence of an infinity of infinite iterations of

$$z \longrightarrow z^2 + c$$

and the distinction between those iterations that result in an orbit that escapes to infinity, and those that do not. So the structure of the geometric object that is the "filled in" circle is still an IIP. So because there are no transformations $\mathfrak{M}$ now has reduced to:

$$\oplus [\odot_c P] \longleftrightarrow \odot_c (\oplus [\odot_c P])$$

where the relation (represented by the arrows in the middle) is the same. This literally means, on the left of the arrows there is are two infinities of instantiations of P, each along one of the axes, with unchanged positions of P along the axes, with also unchanged sizes, forms, and rotations.

On the right, these same infinities of instantiations appear again, infinitely instantiated. All instantiations are exactly the same, in form, size, rotation, and position on the complex plane, and all are instances of the "filled in" circle.

The geometric behaviour of the connected Julia sets when $c \neq 0$ demonstrate that their geometry, which is dependent on c, consists of the combined SS IIP and PS IIP, acting mutually and together.

If the complex plane is a continuum then we cannot accept a discontinuity between $c \neq 0$ and $c = 0$. We should not, therefore, presume a discontinuity in the presence of the SS IIP and PS IIP over the same range of c.

Consequently, we find that the conception of the "filled in" circle as an infinity "points" is challenged by the behaviour of the geometry of the sets themselves, in relation to c. Specifically, we see only at $c = 0$ a form *apparently* accounted for as one object only. At any other value of c close to the origin, the primary object itself, although appearing as a single object, is also a merging of multiple infinities of distinct objects, and it is this that creates the fractal boundary.

Essentially, we can argue, the form of the Julia set at $c = 0$ is still such an object, and its boundary is still essentially "fractal", even though it is "featureless". At higher and higher magnification, a finite section of the boundary has a self similar form, which happens to be a straight line.

Consequently, we can see the "filled in" circle form of the Julia set at $c = 0$ not merely as an infinity of "points", but as infinities of infinities of objects whose essential form is the circle.

The form created by the invariant structure $\mathfrak{M}$:

$$\oplus[\odot_c(T_\odot : P)] \longleftrightarrow \left[\odot_c V_\odot : \{\oplus[\odot_c(T_\odot : P)]\} \right]$$

is dependent on c. So if P_0 is the form of the primary object P when $c = 0$ (the circle), and $\mathfrak{M}_0$ is the specific state of the structure $\mathfrak{M}$ when c $= 0$ and therefore $P = P_0$, then

$$P \longleftrightarrow P_0$$

where the relation is $\mathfrak{M}$.

In other words, the invariant structure $\mathfrak{M}$ of the connected Julia sets, which is always the same no matter what the value of c and hence specific appearance of a particular Julia set, literally creates every connected Julia set form as a transformation of the "filled in" circle that is the Julia set when $c = 0$.

Merging isn't just something that happens in Julia sets when c is inside the main cardioid. Merging happens every time c passes from one bulb of the Mandelbrot set to another of an earlier generation. Conversely, separation happens when c is going the other way.

As the position of c on the complex plane passes outwards from one generation bulb to the next generation bulb, the primary object and every iteration of it, becomes an infinite synapse series.

Coming back into an earlier generation bulb, or into the main cardioid, every infinite synapse series of objects merges into one of the new infinite synapse series iterations of the new primary object, the central synapse series becoming the primary object itself. Well within the main cardioid, there is only the primary object itself. This object consists of the merged infinity of infinite synapse series and their attendant infinite parallel series.

Remember that there are infinitely many bulbs around the boundary of the cardioid, so there is no way for c to exit the Mandelbrot set by continuum change, except by passing through bulbs (or arguably through the limit point of the valley in the main cardioid).

So considering c moving away from the origin, through the main cardioid, and out into descending generations of bulbs, as each synapse is crossed and c enters a new bulb, the process of the separation of each geometric object into an infinite synapse series, is completed.

Because there are infinitely many generations of bulbs, there are infinitely many iterations of the transformation of the geometric objects of the Julia set into another infinite synapse series of self-similar objects. This happens along the two axes, and is what we have symbolised as:

$$\oplus[\odot_c(\mathbf{T}_\odot : \mathbf{P})]$$

- Because of the synapses, of which there are infinitely many, there are already infinitely many synapse series in any given Julia set.

- Each synapse series is itself an infinity of distinct objects.

- Each one of these objects itself "unfolds" a further infinity of objects that it is separated out into, if c moves continuously from one generation bulb to the next (or indeed from the main cardioid to one of the bulbs).

- There are infinitely many times that this can happen, as c moves across the boundary of the Mandelbrot set, because there are infinitely many generations of bulbs.

So we can see here just 4 levels of nested infinities that are manifestly obvious in the Mandelbrot and Julia sets. These are not the only ones, however.

Essentially, when we consider infinities of distinct objects that themselves are expressions of, or indeed are bounded by infinities of, distinct objects that are infinite iterations of these infinities - which is what we see in the Mandelbrot and "filled in" Julia sets, - then we can see a natural extreme to this kind of process. It is an object that consists of infinitely iterated, infinitely nested infinities, in which the nested infinities are themselves are each infinite iterations of the object. We could symbolise this as $\bigcirc$ and could speculate that such objects as the Mandelbrot and Julia sets, and their derivatives such as the so-called "multibrots" are limited, special cases of $\bigcirc$.

Discussion

In the structure $\mathfrak{M}$, we have been considering the objects from which it is composed as the visually discernible *geometric objects* that are empirically apparent in computer investigation of the fractals. However, we should also be aware that this visual evidence is itself only a manifestation of the more general or abstract nature and behaviour of this structure $\mathfrak{M}$, which simply models what arises from the infinite iteration process at the core of which is the recurrence function:

$$z \longrightarrow z^2 + c$$

as a relational structure between IIPs and c values on the complex plane.

We should not forget that this is a two-way relation. In other words the IIP that is used for creating Julia sets, $\mathfrak{J}$:

$$\overline{[\mho\,(\infty)\,\{z \mapsto z^2 + c\}]}^{\uparrow|\checkmark}$$

has a relation with the structure $\mathfrak{M}$:

$$\oplus[\odot_c(\mathrm{T}_\odot : \mathrm{P})] \longleftrightarrow \left[\, \odot_c \mathrm{V}_\odot : \{\oplus[\odot_c(\mathrm{T}_\odot : \mathrm{P})]\} \,\right]$$

that manifests in the appearance and behaviour of the geometry of the fractal object, that is a *two-way relation*. It is just as true to say that the nature and behaviour of $\mathfrak{J}$ arises from $\mathfrak{M}$, as it is to say that the nature and behaviour of $\mathfrak{M}$ arises from $\mathfrak{J}$. The two objects are members of a mutual relation:

$$\mathfrak{M}\longleftrightarrow\mathfrak{J}$$

which is a structure in its own right, and an object that needs to be understood in its proper context.

Ultimately, this is the context of the very means by which we come to be encountering this phenomena in the first place - which is the context of *brain function* and the principle by which it gives rise to the mind and intelligence in which we encounter this.

This is the thing that gives rise to our current *modus operandi* of understanding the nature of numbers and their interrelations and behaviours. Specifically, it is giving rise to the presumption that it is the relations of numbers and their behaviours that results in the other structures we encounter, rather than deeper, currently unseen structures, giving rise to the objects we refer to as numbers.

Page 250

It must be remembered that the complex plane is itself a mathematical construct, an object that we hold in our mind and intelligence. If we consider it as a continuum then it is itself a structure of infinities, that we might endeavour to work with, for example, through the concept of the Riemann sphere, which is yet another object we have instantiated in our intelligence.

Object Theory, in contrast, instantiates the object that we have been calling the *infinite iteration process*, or IIP, in the representation of infinite structures or objects, and it is in this way that the complex plane is represented in $\mathfrak{M}$ in the infinite iteration object $\odot_c$.

Connectedness

A salient feature of the behaviour of the fractal forms as c moves from the origin through the Mandelbrot set and out through the bulbs, is the division of the primary object and its iterated and transformed copies, into infinite synapse series, every time a new bulb is entered. Conversely, as c moves towards the origin, merging, or *conjunction* occurs. What were infinite synapse series of objects, become single objects. When c is in the main cardioid and moves towards the origin, there is only one object, but still with a fractal boundary, that nevertheless could be still regarded as an infinity of objects that are merged with the primary object. Finally, when $c = 0$, all possible objects or merged or conjoined in the "filled in" circle of radius one about the origin.

In this overall scenario of behaviour, the connectedness at synapses appears to have some kind of important relation to the way in which a single object (that may nonetheless include infinite objects along its boundary) can undergo a morphism into an infinity of objects, organised according to an overall principle such as we see in the synapse series IIP.

In terms of Object Theory there is some geometric feature F derived from the initial primary object when $c = 0$, that undergoes a morphism as c changes. This object is sometimes a primary object in a Juliet set geometric form, and sometimes an infinite synapse series of objects. It is also any of the stages in between. This is easily observable in computer imagery.

Page 251

So we could say that this feature or geometric object is literally a *structural function* $\bar{f}$ of the number c. So borrowing conventional symbolism for functions:

$$F = \bar{f}(c)$$

This is another, more specific way of describing how the geometric object F has some relation to its complex plane environment E as:

$$F \longleftrightarrow E$$

We also know that in the process of F undergoing its morphisms, as c moves continuously within the Mandelbrot set towards the origin, F becomes repeatedly - each time a new bulb is entered - composed from the conjoining μ of the previously distinct objects of an infinite synapse series S. So using our earlier notation for IIPs we have that:

$$F \equiv \mu\{\circlearrowleft(\infty)\{S\}\}$$

This new object *itself* whenever c is outside the main cardioid, is always a member of an infinite synapse series of self similar objects. So as c is brought continuously from the boundary of the Mandelbrot set to the origin without leaving the set, there is an infinite iteration process in which the above IIP is infinitely iterated as:

$$F = \circlearrowleft(\infty)\{\mu\{\circlearrowleft(\infty)\{S\}\}\}$$

in which the cycles or iterations of the outer IIP $\circlearrowleft(\infty)$ coincide with the iterations that occur in the Mandelbrot set as the infinite synapse series of bulbs, ending in the main cardioid.

Thus we can say that the "filled in" circle Julia set has the structure given above.

However, this process is also completely reversible, with every infinite synapse series in any connected Julia set, being created from a single object in an "earlier generation" series, associated with c in an earlier generation bulb of the Mandelbrot set. What appears in a later generation series as an infinite synapse series, was in the previous generation a single, distinct object (albeit complete with an infinity of parallel series objects around its boundary, which are other synapse series) in a different synapse series. The change from one to another happens iteratively, the iterations coinciding with the bulbs of the Mandelbrot set.

It is the *synapse series* that are fundamental to the geometric structure of both the connected Julia sets and the Mandelbrot set. The parallel series *are* synapse series, that simply have different spatial *relations* in terms of position on the complex plane, to the main structure of synapse series that spreads outwards from the origin along the two main axes.

The *connectedness* across synapses of the connected Julia sets arises from the fact that the IIP that results in this infinite division of definable single geometric objects within a synapse series, into further infinite synapse series, *is a single* IIP that always incorporates dependence on c. If we take the complex plane to be a continuum, then c which is a *number* contained *within this* IIP can change its value *continuously*, that is, in a way that represents continuum quantities.

Similarly, the connectedness of the Mandelbrot set, quite apart from it being the set of all connected Julia sets, arises from the fact that this feature of *its* geometric structure arises from a *single* IIP that also incorporates c in precisely the same way.

The IIPs have iterations that are distinct objects in their own right, and are part of *iterated structures*. For example, the structure of the Mandelbrot set is a structure of distinct objects that have distinct number objects associated with them, such as the periodicities of the bulbs. The periodicities of the bulbs are inherently related to the infinity of distinct objects that we call the natural numbers.

These distinct, iterated objects *must be* connected, if the c values in the IIPs producing the geometric structures, are to represent continuum quantities. There is no single IIP that will produce values for c that constitute a continuum. We discussed this earlier. This ultimately arises from the fact that *numbers*, and the *continuum*, are

Page 253

themselves two distinct objects. Therefore, the connectedness of the geometric objects is the way that the geometry manifests the relation between the *continuum* object and the *numbers* objects that are subjected to the IIP:

$$\overbrace{[\circlearrowleft (\infty) \{z \mapsto z^2 + c\}]}^{\uparrow|\checkmark}$$

Let's call this IIP **A**. We create it in practice as far as we can, through computer algorithm. Typically, the "decision" object above the horizontal brace is written into the algorithm as the exclusion from the set of any c value (or in the case of Julia set, any initial z value) that results in the orbit of z exceeding the circle of radius to around the origin. (This is because we know that if the orbit exceeds this circle it will definitely escape to infinity).

We cannot fully create **A** in practice, not only because we cannot iterate in practice *infinitely* many times, but also because we cannot test every value of z_0 or c. We might say that this is because they are infinitely many values, but actually, it is because no such thing as "every value" even exist, except as a conception or object in a particular scheme of thought about infinities - namely, the set theory conception.

In Object Theory the infiniteness of c values (or z_0 values) is accounted for in terms of IIPs. Any c value whether irrational or not, is an IIP. We discussed this earlier. The continuum of the complex plane is also an IIP (albeit conceivable as a structure that contains more than one IIP). But no c value IIP is the same object as the complex plane IIP, and there is no single IIP of c values that can be constructed to be the same object as the complex plane continuum.

A little contemplation will show that continuity or connectedness of the connected Julia sets, and of the Mandelbrot set, is possible precisely because *everything* constituting these objects, including numbers, the complex plane continuum, and the IIPs inherent in the geometry and its behaviour, is *related* as a *structure*.

So the nature and behaviour of the fractal geometry of these objects that we otherwise understand as "sets" (the Mandelbrot and the Julia sets) arises in the *relation* that exist in the structure:

$$\circlearrowleft (\infty) \{z \mapsto z^2 + c\} \longleftrightarrow \circlearrowleft (\infty) \{C\}$$

where on the right of the relation arrows is the complex plane continuum. There are two infinities here. There is the infinite iteration of the recurrence function, which consists of numbers and number processes, and then there is the infinite iteration of the complex plane continuum, which is a quantity object without number objects.

We *use* numbers in order to create the sets, which is one way of creating them, through the concept of the "set". The "set" object is nonetheless an object in the mind, and a part of the structure of brain function. It is just one approach.

The fractal objects and their behaviour are not dependent on the set object, because the infinities they manifest are not dependent on the set object. They are first and foremost, IIPs. The fractal objects exist anyway, even without the concept of the set. It is just that we have arrived at our encountering of them, through the concept of the set.

Nevertheless, they are demonstrations of principles in nature, and hence principles in the way our intelligence is working, that are prior to our encountering of these principles as both numbers *and* the observable geometries and geometric behaviours of these fractal objects. These prior principles are *clearly* encountered as IIPs.

Disconnection

It is of course well known that for c values outside the Mandelbrot set the Julia sets are *disconnected*, and are sometimes referred to as Fatou dust.

What happens to the transformation of the "filled in" circle as c passes outwards through the boundary of the Mandelbrot set, to the part of the complex plane outside it?

To pass out of the Mandelbrot set c must pass through a converging infinite series of bulbs, and beyond the convergence limit point.

Conversely, the objects in the synapse series passed through, *never* pass beyond the convergence limit point. Rather, there are infinitely many objects that *converge on* the limit point, through the IIP of *scaling*.

Now *c* occurs as part of the structure $\mathfrak{M}$:

$$\oplus[\odot_c(T_\odot : P)] \longleftrightarrow \left[\odot_c V_\odot : \{\oplus[\odot_c(T_\odot : P)]\} \right]$$

because the transformations applied to the objects always depend on *c*, and on the particular iteration (starting with the primary object) in the infinite iteration process $\odot_c$. However, because *c* is able to pass beyond the synapse series limit point whilst no object instantiated by $\mathfrak{M}$ can do so, *c* must be an object that is distinct and independent from $\mathfrak{M}$, as far as the relation of both *c* and $\mathfrak{M}$ to positions (complex quantities) on the complex plane continuum object are concerned. This is the structural correlation to the concept of *c* being "outside the set".

That is:

$$\mathfrak{M} \longleftrightarrow [\mho (\infty) \{C\}] \longleftrightarrow c$$

must hold as an expansion of the relation:

$$\mathfrak{M} \longleftrightarrow c$$

Remember that the braces around the *C* in the first structure do not denote a *set*, but an object that is an IIP because of the notation on its left. This object in the square brackets is the complex continuum.

If we had said that the complex plane object is a structure of *number objects*, then we would have had:

$$\mathfrak{M} \longleftrightarrow \{C\} \longleftrightarrow c$$

Page 256

where C is this structure of number objects, otherwise usually referred to as the "set" of "all" numbers on the complex plane. This is the structure of the fractal objects conceived as sets, related to the complex plane conceived as numbers, *provided* we *changed the meaning* of $\mathfrak{M}$ and said that it is a symbol standing for the Mandelbrot and connected Julia *sets*, conceived as sets of "points", or "point" or number objects. It does not exist as a structure that we can give any adequate explanation of, if $\mathfrak{M}$ is allowed to remain as it stands. However, if we do change what we mean by $\mathfrak{M}$ then we then completely lose the meaning of $\mathfrak{M}$ as it stands.

The meaning of $\mathfrak{M}$ as it stands is as a descriptor of the IIPs that manifest in the appearance of the synapse series and parallel series of geometric objects that constitute the fractal geometry of the set. It is a descriptor of the actual *fractal nature and behaviour* of the object.

The point here is that this means that if we consider these fractal objects as *number structures*, then we must also consider the complex plane as a set of distinct objects called *numbers*, rather than as a continuum object in its own right. But if we consider the complex continuum as an object in its own right (of continuous complex quantity), then a different picture emerges.

We can begin to see that the way in which the *numbers* on the fractal object behave, and relate to each other, is not just a consequence of *numbers* considered as things or objects that are the fundamentals, but rather, is a consequence of IIPs. In other words, we should consider numbers themselves as IIPs, and their structures of relations as IIPs.

So we can now get back to the question of what happens when c crosses the boundary of the Mandelbrot set, and goes outside the set.

We can consider this in terms of a line that is a continuum, that crosses from inside to outside the set.

The line as values of the quantity c is a continuum that we can represent in our old IIP notation as:

$$\circlearrowleft (\infty) \{c\}$$

or in our new notation as:

$$\odot\{c\}$$

For clarity we will now use the new notation when we are symbolising a continuum quantity, and the old notation when we are symbolising the number.

$\odot\{c\}$ is the IIP of a complex *continuum quantity* that we may then measure with the *number object c*. This quantity, however, is not expressed as a *number* except through the relation:

$$\circlearrowleft\{c\} \longleftrightarrow \odot\{c\} \qquad (1)$$

where now $\circlearrowleft\{c\}$ is the object we call the number on the complex continuum. The fact that we are now using two different symbols for the infinite iterations reminds us that these are not the same iteration process. The one on the left is the IIP of a complex number, whilst the one on the right is the IIP of the complex continuum quantity whose continuum is the line object.

It is the *number* $\circlearrowleft\{c\}$ that is part of the structure of $\mathfrak{M}$ as it is interpreted and understood by number-based mathematics. Here again is the abstract structure of $\mathfrak{M}$:

$$\oplus[\odot_c(T_\odot : P)] \longleftrightarrow \left[\odot_c V_\odot : \{\oplus[\odot_c(T_\odot : P)]\}\right] \qquad (2)$$

In this structure each iteration of $\odot_c$, in all three instances in which it occurs, determines the nature and properties of the transformation objects T and V (which would normally be number matrices), which is why those objects in the structure are given the subscript, to indicate this.

This simply means that on each iteration of the iterator $\odot_c$, the geometric object P may be *transformed differently* by its associated transformation object, and this as we know depends on the quantity c.

On each iteration P undergoes transformations to its spatial position on the plane, its size, and possibly its form and rotational position. The transformation objects apply this transformation.

So the representation of the Mandelbrot set boundary or connected Julia set boundary, in relation to the line crossing the boundary and exiting the set, is represented by substituting (1) in (2), at every instance of $\odot_c$. We then have:

$$\oplus[\langle \circlearrowleft \{c\} \longleftrightarrow \odot \{c\}\rangle (T_\odot : P)] \longleftrightarrow$$
$$[\langle \circlearrowleft \{c\} \longleftrightarrow \odot \{c\}\rangle V_\odot : \{\oplus[\langle \circlearrowleft \{c\} \longleftrightarrow \odot \{c\}\rangle (T_\odot : P)\}]$$

Now for our purposes the fact that this takes place along the two axes is not relevant. So the structure we are interested in is:

$$[\langle \circlearrowleft \{c\} \longleftrightarrow \odot \{c\}\rangle (T_\odot : P)] \longleftrightarrow$$
$$[\langle \circlearrowleft \{c\} \longleftrightarrow \odot \{c\}\rangle V_\odot : \{[\langle \circlearrowleft \{c\} \longleftrightarrow \odot \{c\}\rangle (T_\odot : P)\}]$$

So here we have some continuum objects, and some number objects. Now we can separate out the *number* objects, and any relation such as

$$\circlearrowleft \{c\} \longleftrightarrow \circlearrowleft \{c\}$$

where an object is related to itself, can just be replaced with the object. The transformation objects *are* already *number* objects (because they consist of number structures and relations determining the quantitative (number) properties of the geometric objects. Typically they would be matrices).

So extracting just the number objects and their relations, and remembering the colons are essentially *relations* between transformation objects and the geometric object P, we get:

$$\mho\,\{c\} \longleftrightarrow T_\odot \longleftrightarrow \mho\,\{c\} \longleftrightarrow V_\odot : \mho\,\{c\} \longleftrightarrow T_\odot$$

This is now a representation of part of the structure of $\mathfrak{M}$ considered only from the point of view of numbers. It just represents some generic relations between the transformation objects and the number c. It relates to the way the geometric objects change across the boundary. Let us name this structure Λ. Now anywhere that any of the objects in Λ appear in $\mathfrak{M}$ — which still contains continuum quantities - we can substitute Λ itself, because anywhere there is a relation with any of these objects, there will be a relation with Λ.

So for $\mathfrak{M}$ we now get:

$$\Lambda \longleftrightarrow \odot\,\{c\}\,\Lambda \longleftrightarrow \Lambda \longleftrightarrow \odot\,\{c\}\,\Lambda : (\Lambda \longleftrightarrow \odot\,\{c\}\,\Lambda)$$

as the essential structure of *relations* between the continuum quantities of the line and the numbers on the complex plane that comprise the geometric objects.

In this structure, what is the meaning of this object?

$$\odot\,\{c\}\,\Lambda$$

It means the continuum IIP of the line applied to Λ, which is a structure of number objects. How can a continuum IIP apply to a structure of number objects? The answer is simply that the *iteration* shown here is *not* applying to Λ, it is still applying to the c quantity in the curly brackets. When we write $\odot\,\{c\}$ we have indeed already specified that this notation with the curly brackets *means* an IIP that is a quantity continuum, in this case, the quantity named c. We have however made no such specification for the notation

$$\odot\,\{c\}\,\Lambda$$

This simply is the equivalent a relation we could have written:

$$\odot\,\{c\} \longleftrightarrow \Lambda \quad (3)$$

and we already know what this relation is. It is the application of a number or number structure to a continuum quantity. In this case it is Λ which is the number-definable part of $\mathfrak{M}$, being applied to $\odot\{c\}$ which is the continuum quantity of the line, whose quantities are called c.

What about the structure on the right? It appears as:

$$\odot\{c\}\,\Lambda : (\Lambda \longleftrightarrow \odot\{c\}\,\Lambda)$$

If we substitute (3) from above, we can see it is:

$$[\odot\{c\} \longleftrightarrow \Lambda] : [\Lambda \longleftrightarrow [\odot\{c\} \longleftrightarrow \Lambda]] \quad (4)$$

which is the number-definable part of $\mathfrak{M}$ applied to the continuum quantity c, *transforming* (through the colon relation) *itself*.

This, essentially, is the generic structure of the principle in which the Mandelbrot set or a connected Julia set exists in fractal form in relation to the complex quantity continuum, whose quantities are c.

So in terms of the line which is a continuum of c quantities crossing the boundary of the set, the geometric object that is the Mandelbrot set or the connected Julia set, is essentially a *number structure* that transforms itself iteratively *in relation to* the *continuum quantities* represented by the line. The number structures occur as infinite iterations of transformed geometric objects (infinite fractal forms) precisely because it is only through an *infinite* iteration process that *numbers* can come to represent *continuum* quantities.

Essentially, an *infinitesimal* such as we might employ in the operations of calculus, is literally infinitely small, and yet at the same time not zero, precisely because it exists as an IIP, that is infinite, and does not halt, even when it converges on a finite number. *All* finite numbers are IIPs, both as part of the number generation process, and in many other ways, including that every finite number is the same object as any number IIP that converges on that number.

The transformation here is iterative because of the iterators in the structure of Λ:

$$\circlearrowleft \{c\} \longleftrightarrow T_\odot \longleftrightarrow \circlearrowleft \{c\} \longleftrightarrow V_\odot : \circlearrowleft \{c\} \longleftrightarrow T_\odot$$

These iterators as we can see here are the IIPs of *numbers* that are applied to the continuum c. So the infinitely iterating patterns of form and behaviour that we see in the Mandelbrot and connected Julia sets, along a continuous line, arise from the IIPs of the *numbers* from which the structure of the set objects are composed.

In other words, the form and behaviour of the Mandelbrot and Julia sets, with its infinite fractal, or *iterative* behaviour, is a *manifestation* of the fact that the *numbers* on the complex plane (which in Object Theory are number objects applied to the complex continuum object) are themselves *infinite iteration processes* - IIPs.

In full, what we therefore see in the iterative and fractal form and behaviour of the sets in relation to a continuous line crossing the boundary, which is well represented as geometry in computer imagery, is the *relation* between the IIPs of real and imaginary *numbers* and the IIP of the *continuum*.

Exiting the set

What happens when c *exits* the Mandelbrot set? There is no number that we can say is a "last" number that is *in* the set, adjacent to which is another number that is "out" of the set. That mode of conception is chimerical. It is essentially the same situation with numbers approaching zero.

Gauss hypothesised "hyperreal" numbers to address this issue, but in actuality the issue comes down to IIPs. There is no "last" number that we come to, before zero. The transition from non-zero numbers to zero is exactly analogous to how the line continuum crosses the boundary of the Mandelbrot or connected Julia sets. We simply come up against the fact that numbers are IIPs. Decreasing real numbers applied to a continuum that extends from zero quantity, simply converge on zero, but of course not because they are a geometric series. Rather, they converge on zero because they are IIPs whose individual structures are all instances of a *nested* IIP, as we have already discussed. It is this IIP that constitutes any "number", and it is the IIP that "connects" the *numbers* it constitutes with the *number* zero. The numbers themselves, in contrast, are always distinct objects, both from each other, and from the nested IIP that each number is an instantiation of.

Another way of describing this, as we have already illustrated, is that *quantity* is an object that increases continuously from zero, as a continuum object, whilst *numbers* are objects distinct from the quantity object that can then be applied to the quantity object. This then constitutes a structure. Every number itself is an object distinct from the continuum quantity object. Creating an object called "all the numbers" as a structure whose internal objects are numbers, is not the creation of a continuum, or continuum quantity object. This is still the case even if we want to invent a new kind of "number" object that in plurality exists as distinct objects, and put that together as a structure.

The structure of the continuum quantity is not the same as a structure that emulates it, which then allows us to calculate. This is rather like the way in which digital representations of continuum phenomena may work perfectly adequately for us, but that does not mean that continuum phenomena is working digitally.

The line crossing the boundary is a continuum object. As such, it is distinct from any number object that may be used to represent any quantity of that continuum. So where the line crosses the boundary, is a structure of three kinds of object:

The first is the continuum object of the line, as a representative of part of the continuum quantity of the complex plane, whose quantities we label as c.

The second is the number objects that are used to represent the quantity c.

The third is the object we call the Mandelbrot set, which can be understood in a number of different ways. Firstly, of course, it can be understood through the concept of the "set" of numbers. This concept is an object in its own right. Secondly, if we wanted to, we could understand it as a continuum object. Thirdly, as we have been doing, we can understand it as a geometric object on the geometric representation of the complex plane object.

There is, however, a fourth way. This is to consider the Mandelbrot set *as* the relation between the number objects applied to the complex plane continuum, and the complex plane continuum quantity itself. This is basically what we have already just derived, above.

So we are saying this structure is:

$$\odot\{c\} \longleftrightarrow M \longleftrightarrow \circlearrowleft\{c\}$$

where M is some suitable descriptor of the structure of the Mandelbrot set, as a structure of IIPs involving complex quantities.

So this is an alternative, simple way of writing that the Mandelbrot set object literally stands as a *relation* between the number objects applied to the complex plane (the IIP shown on the right) and the complex plain continuum quantity itself (the IIP shown on the left).

Dust

Disconnected Julia sets, where the value of c is outside the Mandelbrot set, are sometimes referred to as fractal "dust" or "Cantor dust" or "Fatou Dust". This depends on which structure or system of concept structures is being used to describe the phenomenon.

It must always be remembered that the predisposition of *naive realism* is to presume that the concept structures being used in mathematics, because they are widely used, and are things that been learned about, are *genuinely objective* (by the criterion of Object Theory), and exist independently of the mind or network of minds that are using them. This is simply not necessarily the case.

It *is* the case that all modern mathematics consist of inter-related concept structures, some of which translate into actual structures of relations between *numbers*, and some of which do not. But furthermore, and more importantly, those structures that do translate to structures of relations between numbers, are still just structures that fulfil that role, without necessarily being in themselves genuinely objective.

For example, if we take the case of the behaviour of the relation or function $y = 1/x$ to be genuinely objective as a relation of continuous quantities, then we can still model it using discrete mathematics or digital means, which are genuinely objective, but not, in themselves, the same thing as that relation of continuous quantities.

Mathematics deals with in *representations*. Ultimately, even that which we consider to be *genuinely objective*, because by the criterion of Object Theory we see it as not dependent on any particular individual mind or network of minds, is still a *representation* of something that is *not yet understood*. It is not yet understood, because it is not yet understood how brain function equates to the intelligence and experience of being and world in which this comprehension is arising in the first instance.

The concept of "dust" is such an object. It only relates to numbers and number relations, if our conception of it consists of the appropriate structure.

Cantor dust is often described, if we are using the terminology of Object Theory, as the IIP:

$$\circlearrowleft (\infty) \{L\}$$

where L is the process of removing the middle third of any line in the object domain to which the IIP is applied. Obviously, beginning with one line, removing the middle third leaves two lines. So on the next iteration of the IIP the middle third of each of these lines removed. And so on.

It is often conceived that "when the process is complete" then what we are left with is "Cantor dust". This conception is chimerical. The reason is that the IIP does not halt. So the concept of "when the process is complete" is false.

This kind of thinking arises from *naive realism*. It arises from the idea that there is something separate from the processes of mind and thinking, that the processes of mind and thinking are being applied to. In fact, all there is, is the processes of mind and thinking.

If Cantor dust is to bear any relation to numbers and number structures, then we must have that:

$$\circlearrowleft (\infty) \{L\} \longleftrightarrow R$$

where R is an object that stands for real numbers. However, the IIP on the left, for the process of producing Cantor dust, is infinite and does not halt. R cannot be a single IIP for "all" real numbers, because there is no such thing (as we have previously shown).

The concept of Cantor dust as an object, applied to the complex plane, is a specific instantiation of a more general structure we can call D, that would be an IIP for producing an infinity of distinct numbers that are *disconnected*, but are nonetheless applied to the complex plane continuum object. What does this *disconnected* mean? It means there is no connection or *continuum* between one number and another.

Essentially, any complex number we are talking about is a distinct state of the IIP for a complex number. Let's just talk about the real number part of that, since this will suffice.

Every real number is a distinct state of the real number IIP for the given number-base. Any real number is a specific instantiation of that IIP. We can consider it halted, if we wish, in the case of a rational number. But it can always be considered unhalted, in the way we have already described. Even a rational number may be considered as the unhalted IIP because it can be considered as having an infinity of zeros after the last digit. The iterations of the "inner" IIPs (or in our earlier analogy the wheels of the machine) may be halted, but the "outer" IIP, the process that infinitely produces wheels, is still unhalted. So the overall IIP is an unhalted IIP.

Each unhalted state of the IIP is an object in its own right. This is the real number. It is inherently distinct from any other number, even one that may be taken to represent the same continuum quantity (such as in the case of 1 and $0.\dot{9}$). Thus any real number is already inherently distinct *and disconnected* from any other real number, unless it is *connected via* another object of another kind.

We can got on replacing the relation $\longleftrightarrow$ between two distinct real numbers in the structure:

$$a \longleftrightarrow b$$

with another real number *ad infinitum*:

$$a \longleftrightarrow b$$
$$a \longleftrightarrow c \longleftrightarrow b$$
$$a \longleftrightarrow d \longleftrightarrow c$$
$$a \longleftrightarrow e \longleftrightarrow d$$

This in itself is an IIP. There is always a self-similar form in the structure itself. So essentially, real numbers, or complex numbers for that matter, are already a form of fractal *dust*. The introduction of the concept of hyperreal numbers, for example, is completely unnecessary, because the IIP for a real number is already capable of infinite nested iteration.

The only way we define real numbers not to be dust, or to be a continuum, is by creating structures *defined* in such a way that we can *say* that they are not dust, and/or are a continuum. But then we may run into what are essentially paradoxes, such as the notion of dust that is a continuum, which it is possible to arrive at as a conclusion, through certain means of mathematical analysis.

The *complex continuum object* in contrast is *by definition* a continuum quantity, and is not dust.

18: The Geometric Exit

Geometrically, the continuum quantity c along the line must leave the Mandelbrot set either from a bulb, or arguably from an "alternative" such as the limit point of the valley in the main cardioid. However, with a little extra work we may show that even these "alternatives" are always IIPs.

The geometric objects in the Mandelbrot set that reiterate the shape of the main geometric object, are often called a "minibrots". Again, there is no "last" minibrot on the far left of the "filament" on the negative real axis. Rather, there are infinitely many, and if we were to engage in a process of looking for the "last minibrot", we would be caught in a process of infinite zooming, in order to try to find it, but we would never find it, because the nature of the geometric object we are examining, is an IIP. It does not halt. However, because of the transformations applied by the IIP to sizes and positions of the geometric objects, the synapse series does converge on a finite number or quantity on the complex plane real axis.

On the first image of the following pages we see the extreme end or convergence point of the Mandelbrot set on the real axis, on the negative or left side. Along this "filament" are infinitely many iterations of "minibrots". The principle of the infinite iteration process continues everywhere.

The synapse series formations that occur in the "filaments" of the Mandelbrot set in general, invite further study. They appear to exhibit nested synapse series, together with the principle of synapses themselves consisting of synapse series objects - a feature found throughout the "filament" structures of the Mandelbrot set. There is potentially greater insight to be gained from this.

Above: The far left extreme (on the negative real axis) of the Mandelbrot set. The large bulbs shown belong to a "minibrot" along the "filament" extending along the negative real axis, not visible when the entire set is visible.

The synapse sequence extends to the "point" of the object, on its left, but it is not only one synapse series that is present. There are "nested" series, just as the parallel series formations consist of infinitely many "nested" series.

Below: The last *findable* "minibrot" along the negative axis at a zoom factor of 10^13 and a 10,000 iterations escape time algorithm. There is no "last" object in the actual series, because it is an IIP. What is visually "findable" depends on computing power.

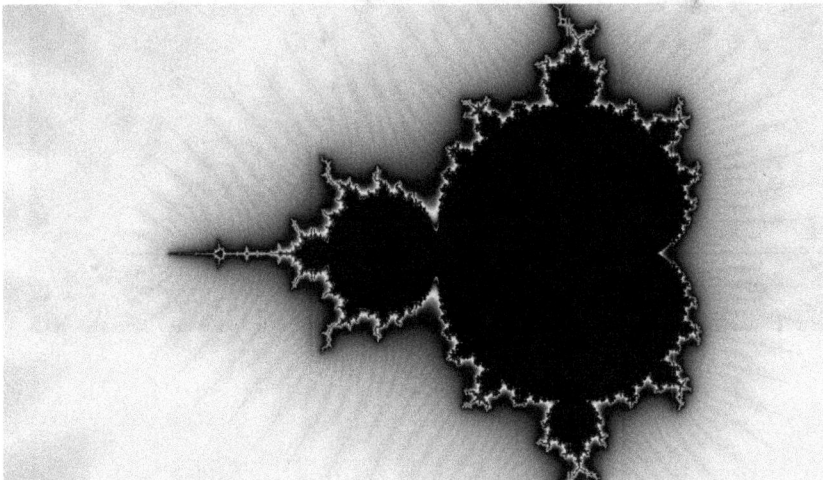

Above: A "final" connected Julia set form when c is very close to the edge of the Mandelbrot set but still within it. As *c* continues to move towards exiting the set but remains within it, the form of the Julia set remains stable.

Below: the same Julia said, but zoomed in. The dark objects near the centre of the image are the connected part of the set, which appears in the first image on the next page, zoomed in more.

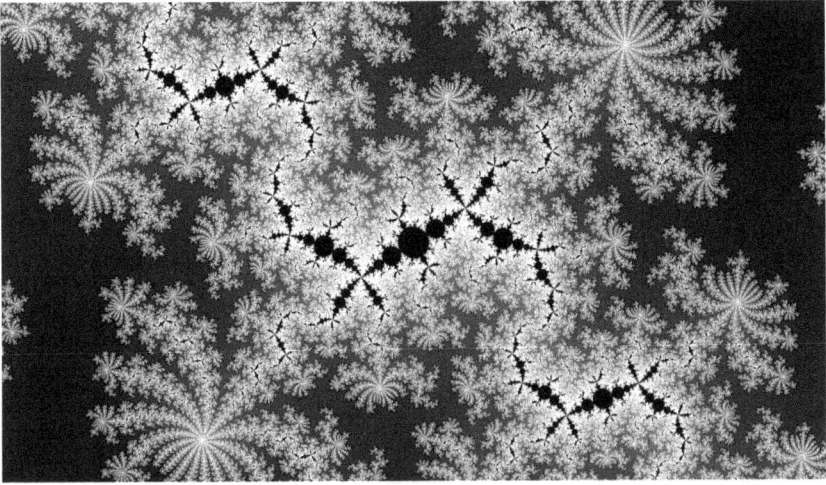

Above and below: the previous image, zoomed in more, as c is close to exiting the Mandelbrot set. The form here is of the synapse series of objects midway through the transformation process as c moves from one generation bulb in the Mandelbrot set, to the next, whilst moving towards exiting the set. There are infinitely many bulbs, each generation a fraction of the size of the previous generation bulb. As c moves into each further generation bulb, each individual object in these series in the Julia set, transforms into an infinite synapse series of self-similar form. This process repeats infinitely.

The overall form of the set is stable as the actual changes in the *number value* of c from one bulb to the next is extremely small.

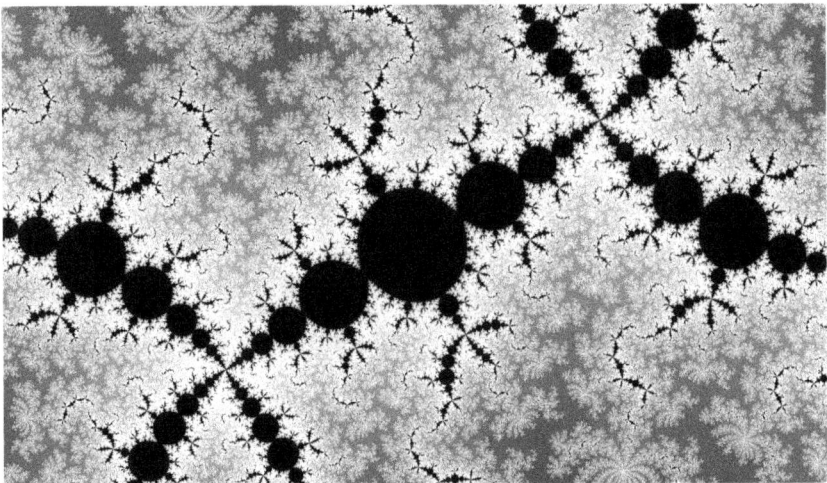

In the previous images we see some snapshots of the form of the Julia set as c is about to exit the Mandelbrot set. The value of c moving towards exiting the Mandelbrot set passes through cycles of position relative to the bulbs that it passes through. This process is an IIP. Passing through a bulb is the thing that is being iterated.

In the Julia set, the corresponding iterations are the transformation in the shape of the individual synapse objects, which takes place psychically with each iteration, and once in each iteration every individual synapse series object breaks into a new, infinite synapse series. This too, is an IIP, synchronised with the IIP of the value of c passing through the bulbs of the Mandelbrot set.

Each IIP itself does not halt. Neither does the complex continuum quantity of the line crossing the boundary of the Mandelbrot set. These do not halt, in themselves, because they are IIPs. Nevertheless, we can, if we choose, *apply* a halt to any IIP.

If we do this to the IIP that is the process of c values crossing bulbs in the Mandelbrot set, towards existing the set, then we have a specific complex number that we stop at. There will also be a connected Julia set for that value of c. It will be a halt-point somewhere in the IIP of transformations of shape and size of the set.

Similarly, we can apply a halt to the IIP for the line continuum, rather than letting it be infinite. Then we have a line of finite length.

The actual crossing of c values in the line continuum out of the boundary of the Mandelbrot set, into the zone outside the set, takes place in the *relation* between the IIP of the line and the IIPs of the geometry of the Mandelbrot set. Essentially, we can ignore the parallel series IIP, and just consider the synapse series IIP.

Let us call the synapse series IIP of the Mandelbrot set, M.

Let us call the synapse series of the corresponding Julia set for any c value, J.

Let us call the IIP of the complex continuum quantity represented by the line, L.

Let us call the c values as numbers, c.

Then we have the relations, and the structure:

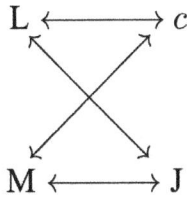

$$L \longleftrightarrow c$$
$$M \longleftrightarrow J$$

L $\longleftrightarrow$ c is the relation between the continuum quantities (on the complex plane) that the line represents, and the number c applied to each quantity.

M $\longleftrightarrow$ J is the relation between the IIPs present in the Mandelbrot set where the line is, and the IIPs of the Julia sets corresponding to the values of c on the line.

The relations L $\longleftrightarrow$ M and c $\longleftrightarrow$ J are not so much of interest here as we are already familiar with them.

What is of interest are the relations L $\longleftrightarrow$ J and M $\longleftrightarrow$ c.

The relation L $\longleftrightarrow$ J is of interest because somewhere in this relation J stops corresponding to a connected Julia set, and starts corresponding to a disconnected Julia set, or "dust".

The relation M $\longleftrightarrow$ J is of interest, because as we have shown, and there is a way in which M and J are both the same structure. It is only the relations of specific parameters within the structures, that result in the differences. This of course also manifests in the fact that both geometric objects can be generated using the same recurrence equation.

The relation M $\longleftrightarrow$ c is the relation between the IIP of the number c and the IIPs of the Mandelbrot set geometric object.

Let's look at the relation L $\longleftrightarrow$ J. This relation also exists as the object M. We can replace the relation arrows with M because it already exists with a relation to both L and J. This is the normal procedure of replacing relation arrows with an object. So we are looking at:

$$L \longleftrightarrow M \longleftrightarrow J$$

M is the IIPs of the Mandelbrot set geometric object - its synapse and parallel series IIPs.

L is the IIP of the complex continuum quantity represented by the line.

J is the IIPs of the Julia set geometric object. These are structurally the same as those of M - they are synapse series and parallel series IIPs, but they have different parameter values.

So we can see that what is of interest is the relation between the complex continuum quantity represented by the line, or L, which is an IIP, and the synapse and parallel series IIP structure that exists in both geometric objects, the Mandelbrot set, and the Julia set.

We already showed a structure for the IIP of M and J. In the previous chapter, as structure (4) we showed that the simplified structure of the geometry of both the Mandelbrot set and a connected Julia set is:

$$[\odot \{c\} \longleftrightarrow \Lambda] : [\Lambda \longleftrightarrow [\odot \{c\} \longleftrightarrow \Lambda]]$$

where Λ is the number definable part of the structure of the geometric object, and $\odot \{c\}$ is the IIP for the complex quantity continuum - which in this case, is the line. The relation

$$\odot \{c\} \longleftrightarrow \Lambda$$

here, and when it is written laterally inverted, is the application of the number definable part of the structure to the complex continuum quantity that it relates to and appears on. The colon means that the structure on the left of the colon transforms (as scaling, position, and rotation, of geometric objects) the structure on the right. But remember that the object Λ as the number definable part of the geometric structure itself also contains transformations.

So looking from left to right, we can see that the first object in the square brackets is the relation of the number definable part of the geometric structure to the complex continuum, which in this case, is the line.

This then transforms an instance of itself in the object on the right of the colon, transforming at the same time the first Λ after the colon and square bracket. This Λ is another identical instance of the number definable part of the geometric structure, but here, it is not applied to the continuum quantity $\odot\{c\}$ - the line considered as a continuum.

The structure on the left of the colon is the structure of the geometric object is considered as complex numbers applied to the continuum quantity $\odot\{c\}$ which in this case is the line.

The structure on the right of the colon which is being transformed by these numbers, is the transformation of *both* these numbers *and* their relationship with the structure of the geometric object considered as numbers alone, independent of the continuum.

Because these are all IIPs what this is essentially showing is that the geometric structure is infinite iterations of transformations of itself, which is its fractal nature.

However, *what gets transformed* is not just the geometric structure in relation to the continuum, which is:

$$\odot\{c\} \longleftrightarrow \Lambda$$

This is an object in which we can consider Λ as a number-structure that can be treated as a continuum because all its numbers are related to the continuum.

In addition, what also gets transformed is Λ *as just a number structure alone* with no relation to the continuum, but *together with* its relation to Λ considered as a continuum. This is the part that is:

$$\Lambda \longleftrightarrow [\odot\{c\} \longleftrightarrow \Lambda]$$

So the fractal behaviour of Λ, *in its relation to the continuum* (or the line) $\odot \{c\}$ *is occurring in this context.*

So the relation between the *numbers* in the Λ structure (which is the geometric object) and the continuum, stands out as part of the infinite iteration and transformation processes that constitute the geometric objects.

So essentially, what happens as number values along the line (which is the continuum $\odot \{c\} \longleftrightarrow \Lambda$) are pursued further and further towards the edge of the geometric object, is that the self transformation of the "continuum-like" geometric object is always related to the non "continuum like" geometric object.

Of course, in computer imagery we just see a "geometric object". The "continuum like" object and the non "continuum-like" object objects in our intelligence that ordinarily, would depend on which approach we are taking to analyse this phenomenon.

The non "continuum-like" geometric object Λ is the geometric object considered as numbers. Of course, in another approach the argument would be that the numbers *are* a continuum. But in Object Theory this is not the case.

So here we are looking at, *in essence* as the mechanism behind the object being transformed with values of c:

$$\circlearrowleft \{c\} \longleftrightarrow [\odot \{c\} \longleftrightarrow \circlearrowleft \{c\}]$$

where $\circlearrowleft \{c\}$ is the IIP for a complex number, and $\odot \{c\}$ is the IIP for the continuum. The structure in the square brackets is also an instance of the structure that is transforming this whole structure.

In terms of numbers and the continuum, alone, this is describing essentially what happens to this Julia set structure that we saw above, occurring just before c exits the Mandelbrot set:

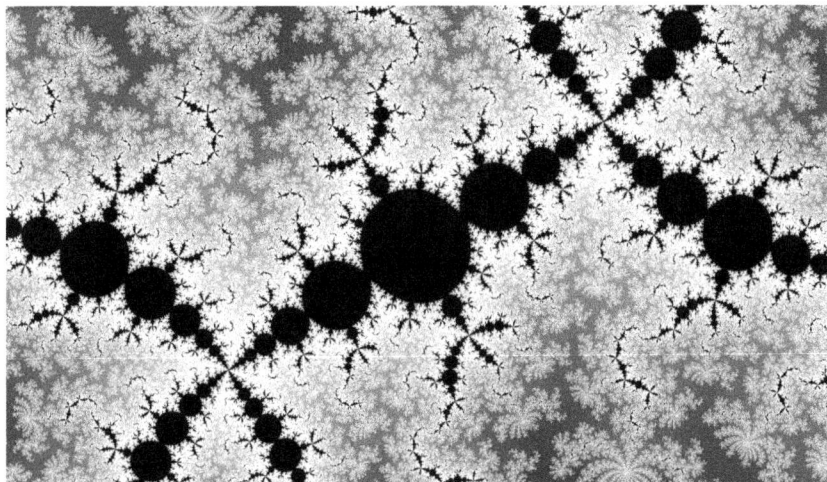

The scaling factor is contained within the object:

$$\odot\{c\} \longleftrightarrow \Lambda$$

which stripped down to essentials is:

$$\odot\{c\} \longleftrightarrow \circlearrowleft\{c\}$$

and its outer iterations (inside which this is nested) that occur as c moves from bulb to bulb in the Mandelbrot set, are synchronised with the outer iterations of *the other instance of this structure* in the Mandelbrot set geometric object. This process is an IIP and so the process of the transformation of the object illustrated above into objects of smaller and smaller scale, is infinite. Nowhere in this process does the geometric object become disconnected.

Remember that there *is no* outer IIP for the complex number IIPs $\circlearrowleft\{c\}$ that we can create and then apply to the continuum $\odot\{c\}$, such that we actually create a continuum of numbers.

Only the continuum, the line of complex continuum quantities, crosses the boundary of the Mandelbrot or connected set, as a continuum. The geometric structure of the set itself consists in its bare essentials, as the structure we have been looking at, of *relations between* between complex numbers and the complex quantity continuum.

Thus, there is no "point" at which the line exits the set. Because a "point" in this context is a *number*. Nevertheless, the line, the continuum, exits the set, which is the structure of *relations between* the continuum and the numbers.

Where the line continuum has already exited the set, it is no longer a part of that structure. We are then left either just with a structure of distinct instances of ↺ $\{c\}$, or no structure at all. In the first case, it is the "dust" of a disconnected Julia set - distinct complex number objects ↺ $\{c\}$ that cannot be perfectly identified as number objects other then IIPs because (actually like all numbers) they *are* IIPs.

In the second case, in which there is no structure at all, it is the area outside the Mandelbrot set, which is the complex *continuum*, and indeed is a continuum object, but is not a part of the structure of the Mandelbrot set object.

Thus we see that the relation between the *continuum* (object) and the *complex number* (object) is *demonstrated* by the Mandelbrot and Julia set objects, to be exhibited in the *structure of relations* between *infinite iteration processes*.

The implication is that everything can be described in terms of structures of relations between infinite iteration processes, an approach which provides a different method of handling infinities.